DARK MATTERS AND PROPHESY: SECRETS OF THE OUIJA BOARD
(UNVEILED AND EXPOSED)

BY JEFERSON DE SOUZA

MY GREATEST GRATITUDE TO ALL
THE ONES THAT MADE THIS BOOK
POSSIBLE. I DEDICATE IT TO MY
FAMILY AND THESE INCREDIBLE
PEOPLE THAT MADE IT TO BECOME
REAL.
MY SPECIAL THANKS TO:
CARLOS E. ALBUQUERQUE,
MARTHA M. ALBUQUERQUE,
CARLOS A. KUMMER
AND CASPER CENDRE,
JEFERSON DE SOUZA – THE AUTHOR

THE OUIJA BOARD

Spiritualists created the apparatus, but businessmen popularized it. The Ouija: interesting and mysterious... Is it simply a novelty, or is there a sinister side of it as well? Join us on an effort to understand this peculiar device. From its beginnings in a faraway post to its rebirth in modern times, the Ouija has traveled a long way. Is it a myth or an unknown fact?

You, reader, will be the ultimate judge of that.

"I CANNOT CAUSE LIGHT; THE MOST
I CAN DO IS TRY TO PUT MYSELF IN
THE PATH OF ITS BEAM."
ANNIE DILLARD

"THERE IS NO SUPERNATURAL, BUT
ONLY THE NATURAL YET TO BE
UNDERSTOOD."
JEFERSON DE SOUZA

Jeferson de Souza and his research team have revealed what was previously hidden about the origins of the Ouija... Through countless hours of international investigation, Souza's team found evidence that suggests a deep history of the U.S. Government's involvement in paranormal experimentation.

In order to collect what is assembled here for you, they had to go above and beyond the call of duty and collaborate with researchers far and wide. To examine the complexities involved, the secret history of the U.S. Government's investigations into extrasensory perception and psychokinesis must first be revealed.

For almost half a century, the U.S. government has researched extrasensory perception at length and without publishing findings. Other countries, including Russia, were leaders in paranormal research since the beginning of the original U.S.S.R. This examination into those theories and the organization of the team's analysis brings to light what government agencies have spent years of research and countless tax dollars on. This book serves to show the people what the Powers That Be have already known for generations - that paranormal activity exists.

Book editing by Casper Cendre, Max Durbin, and Bread Tarleton

INDEX

DISCLAIMER

It has been said and spread to certain circles the following claims: DARPA (Defense Advanced Research Projects Agency) of Arlington, Virginia had allegedly dedicated certain initiatives regarding engineering programs' physics, quantum computing interface, and quantum mechanics in order to produce devices that operate in unusual environments (and even sometimes in non-environments). The author will not confirm nor deny such claims.

"With Knowledge comes more doubt."
 Johann Wolfgang Von Goethe (1749-1832)

FOREWORD

We travel abroad to admire the heights of mountains, the mighty waves of the sea, the power of tides, the majesty of rivers, the greatness of the oceans, and the circuits of the stars. Yet we pass over the mystery of ourselves without a second thought. Most of us walk over the dimly lit streets of our lives encircled by the distractions around us. Even when we are surrounded by people, we barely notice a soul in sight. We go on without making much sense of our very existence. In the darkness of such oblivion, all we can do is wait for another day.

Humans are an incredibly complex sentient species. Millions of us choose to live in cities of varying sizes. We are solitary beings amongst a multitude that hates to be together, but fears being alone. Humans are powerful pits of emotions, waiting to erupt. We are paradoxical entities made of the worst and best of all matter. Is that not what lies at the core of each of us?

We are shrouded by thin veneers of societal expectations that cover the deeply buried subconscious of basic instincts and primal emotions. Most of us live half-lives, operating upon the structure of routine, as if we were living automatons of the ordinary and mundane. Yet we remain oblivious to all of it, as life passes us by.

Only few of us have the courage to see differently. The capability to break free; to reach past what is expected of us and seek the unimaginable: like clumps of charcoal becoming polished diamonds. We can evolve from our apathy and reshape ourselves into what we were meant to be.

There will always be an army of visionaries and futurists offering thoughts and opinions regarding this, who will go to great lengths to prove their viewpoints. No method is better than the next, as the results are the true compass towards their effectiveness.

The paranormal attracts most people like moths to a flame. Our curiosities push our minds past the realm of ordinary existence. We crave wonders and terrors, the uncanny and unusual—all to feel alive but for a mere moment. The Ouija board was meant to be seen as just that: a toy to be played with, a feeling to be chased. Soon after, it became a tool for spiritualists. Now we know it is not merely a toy or a tool: it is a key. A key to amazement and terror alike, opening doors to elsewhere and even the fearful nowhere. What lies on the other side is all up to the cosmos. Be aware of the choices you make.

The Ouija board waits patiently for your attention.

SECRETS OF OUIJA BOARD

3

PROLOGUE

Ouija. How much do you really know about it?

In its debut as a novelty item in the early 19th century, it appears in the papers as "OUIJA: THE WONDER TALKING BOARD." It was originally boasted as a toy and novelty item at a shop in Pittsburgh, Pennsylvania. The Ouija's creators marketed it as a device of magical capabilities that could answer questions about the past, present, and future with marvelous accuracy. They promised never-failing amusement for people of all classes. An object of the known and the unknown, the material and the immaterial. Between the successful marketing and the mystical draw of the Ouija itself, it has remained extremely popular to this day. But is it a simple toy, an instrument of spiritualism, or something far more nefarious and vicious?

The Ouija (pronounced wee-ja) board is a product of the arcane, and intrigues the curiosity of people in ways beyond the natural, beyond the conventional, beyond the tedious sameness of their mundane lives.

The story of the Ouija started millennia ago, in what is now known as Iraq. The grand vizier, Alab Al-Calib ibn al-Dalab, was a scholar of the Ziggurat of Ur (circa 4000 BCE), previously a city in ancient Mesopotamia. People sought out the Ziggurat ("temple whose foundation creates aura") for both physical and spiritual nourishment. During his studies there, Alab Al-Calib ibn al-Dalab devised an instrument of communication with the ethereal world. He called this instrument the "eyes of the heavens," or the Tudla. It was said that it opened the doors of many worlds, and the only thing it required was a living key.

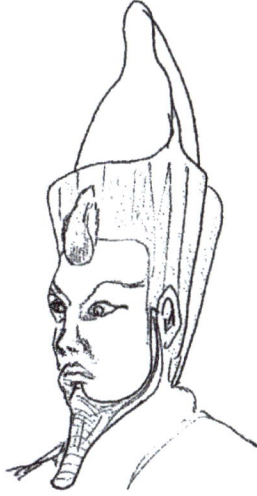

ALAB AL-CALIB IBN AL-DALAB

THE TUDLA

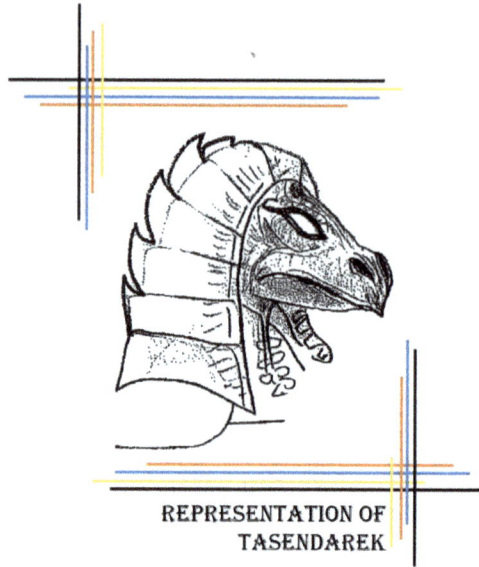

REPRESENTATION OF TASENDAREK

The Akkadians inherited the knowledge of the Tudla. Tasendarek, the reptilian "Godling of the Whispers," perfected the Tudla, naming his new creation, the Maradah. The Maradah was a far more powerful "key to the underworld." He ventured on a quest to lower Egypt, once the land of the Nubians, to bring the key to the underworld realms to Hassin Alab ibn Faruk.

In Egypt, Hassin Alab ibn Faruk was guided by the god, Anubis, to melt the Maradah and remake it into the Udjatti, "the Eye of the Sun and the Eye of the Moon" or "the Eyes of Heaven."

For decades it remained just there as the musings of a scholar and king. That is, until Ali Alhazen took upon its benefits to find great success as a chemist, mathematician, physicist, poet, and, most of all, an optical scientist. He used the Eyes of Heaven to "see far beyond," but perhaps the knowledge was too much for him. For many years he was considered insane by his colleagues and spent most of his life as a housebound recluse, writing and studying things that the "voices from beyond" allowed him to know. During that time, he wrote his seven-volume Book of Optics which changed the way people understood eyesight and later influenced scientists such as Roger Bacon and Johannes Kepler.

THE MARADAH

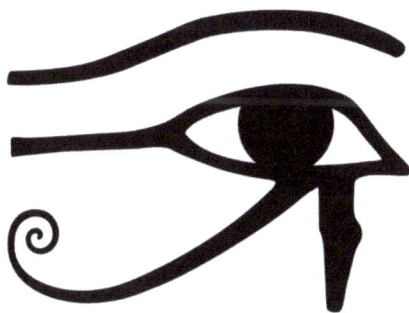

THE UDJATTI

ALI AL-HASAN IBN AL-HAYTAM
(ALHAZEN)

FIRST DARK ROOM - PHOTOGRAPH

8

Alhazen's work focused on the nature of light. For instance, he built a system to study the visual perception of images and the mechanisms of the eye. His studies on lenses led to the development and production of the first set of eyeglasses and eventually the prototypes for what would become telescopes and microscopes. He identified the basic principles of modern photography, which was unknown outside of China at this time. By combining two or more lenses he created prisms, using them to form an image for viewing or photographing. In doing so, he introduced the camera obscura to the western world.

Alhazen, born in the year 965 AD, was perhaps one of the first scientists in history (due to the rigorous methods of experimentation he used), and yet, he also had a personal belief in the Djinn (a so called "spirit" in Muslim legend often capable of assuming human or animal form and exercising supernatural influence over people). Secretly, he was said to receive instructions of "the voices from beyond" through the Udjatti by one or more Djinn. It is believed that Alhazen concealed his use of the Udjatti from the mullahs (from Arabic Mawlā, meaning master or [close] friends) because he wanted to keep his dealings with the Udjatti and the Djinn away from them.

Alhazen, the governor of Cairo, was obsessed with the idea of the Eye of Horus ("Udjatti,") especially after studying the many sacred names of gods. He translated Hebrew characters and the many versions of the Yawe (the vowels here being for the benefit of the reader as Hebrew has only consonants) to have a single meaning: "he made to become." One of the names of a God that took his interest was Yihowa (ee-oh-wa) and Alhazen felt a special significance for it that would find its way into his work.

Due to differences with the caliph Al-Hakin, Alhazen had to trick the royal family and his peers into believing that he had become truly mad in order to be left alone. So, for eleven years he deceived them, staying locked away in his home as he continued on secretly with his experiments guided by the Udjatti. In this time he trained a selected group of young men to continue his legacy after his death. And when caliph Al-Hakin died, Alhazen expanded his resources and trained his protégées with the techniques of the Udjatti. He died soon after the caliph.

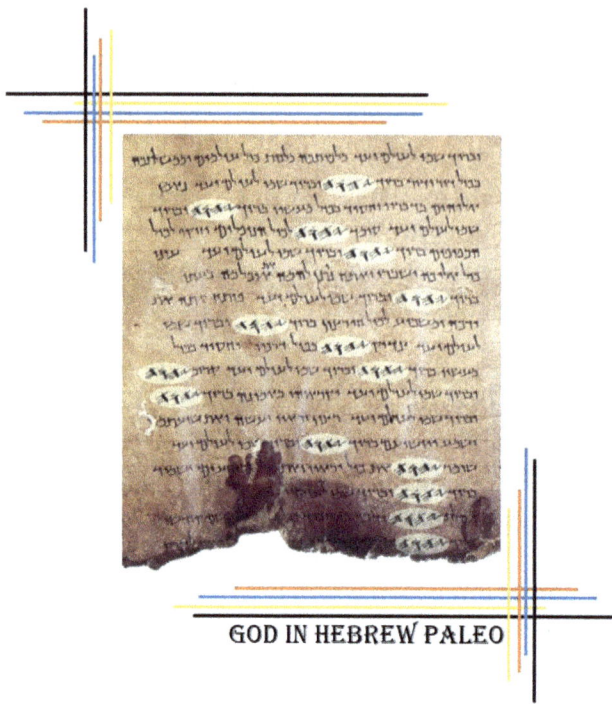

GOD IN HEBREW PALEO

Faruk Adib abn Duzzar was the grandson of Farik Alab abn Duzzar, the greatest of Alhazen's apprentices. At his thirtieth summer, he was told to go to the newly-constructed city of Marrakesh, in the foothills of the Atlas Mountains. He was to bring the Udjatti to a necromancer there, who was known to be at a certain obscure bazaar of the Moroccan city.

Marrakesh is a city of west-central Morocco, sitting at the very foothills of the fabled Atlas Mountains. It was founded in 1062 and soon became a commercial center and a popular hub noted for its remarkable leatherwork. It was also secretly famed for its unique fetishes and the findings of rare objects of questionable use.

Once there, the necromancer consulted the ethereans and saw the next evolution for the Udjatti. Another place in time was given and he was instructed to take the hide of a fierce bull and carve the "Eyes of Heaven" and an alphabet upon it. If these directions were followed, the hide would be a key to the etheric worlds, giving voice to the entities from far beyond.

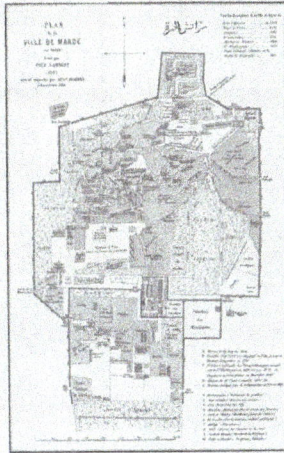

MOROCCO 1950

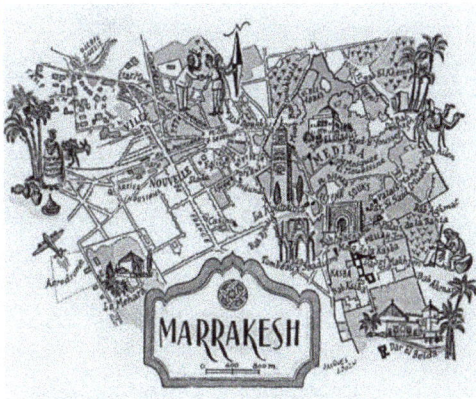

OLD MARRAKESH MAP, MAROCCO

11

The necromancer worked a long time to prepare the special leatherwork that would depict the ultimate representation of the "Eyes of Heaven."

Thus came the "Oudja," the necromancer's representation of the "Eyes of Heaven" in his leatherwork display. It is rumored to have writings on its back with incantations made from the blood of the sacrificed Faruq Adib abn Duzzar, the mad Arab.

They named the leatherwork "Oujda" after the fabled Moroccan city, and it would remain there until the 1800's. These are the known origins of the Ouija.

CHAPTER 01

THE RISE OF SPIRITUALISM

Apart from the witch burnings in Salem, Colonial America is thought to offer little in the way of occult history. However, Americans have been consulting with ritual magic since early times. The night following George Washington's oath of office at Federal Hall in New York City, forty deluded souls gathered in a large circle in an open field near Morristown, New Jersey, to wait for instructions from spirits about the location of caches of treasure in Schooley's Mountain.

Even when such practices were considered to be demonic, some faithful Bible followers also believed in Bibliomancy, a practice that (according to them) was raided by the "will of the Holy Ghost." The practice came from the Puritans overseas and used a ritual called Gospel Cleromancy that worked by means of "spiritual guidance." The ritual involved the questioner (a person chosen to ask their questions to the spirits) wearing a blindfold, opening the Bible to a random page, and selecting a verse that would, supposedly, be the answer to the question.

In the mid-1800's, spiritualism was seen by many as a pursuit for drunks and the superstitious. Still, it managed to sneak its way

into the minds of many Americans and by the end of the 19th century, spiritualism was on the rise. The public's renewed interest in alchemy, astrology, ghosts, spells, and magical cures suggested that they had come a long way since the Salem witch trials.

Some of the most interesting accounts on the early Spiritism movement came from an American journalist named Elizabeth Cochrane Seaman (better known as her pen name, Nellie Bly). She was known for her muckraking articles in the publication The New York World, particularly her undercover investigation and exposé on the abhorrent conditions at the Women's Lunatic Asylum on Blackwell's Island (*Ten Days in a Mad-House*) She also wrote a famous account of her 72-day journey around the world. However, very few had access to her personal accounts of a growing trend all over Europe: the séance.

The origins of the séance are surprisingly mundane, given the mysterious atmosphere associated with the word. It comes from the French term for "seat session," from old French "seole" meaning "to sit." In French, as in English, the words came to be used specifically for a meeting of people to receive spiritualistic messages (a séance was first recorded in English in 1845).

As Spiritism goes, its beginnings start way, way before that.

For instance, in Oppenheim, Germany in 1620, a goods-trader moved to a new house. One night he heard some strange noises from within the walls, though he knew he was alone there. He bought his house from his sister, the widow of Humbert de Birk, who died in the house one week before. The new master of the house thought that the raps he heard could be his late brother-in-law trying to communicate with him. So, he asked the ethereal presence, "If you are Humbert, tap on the wall three times." He heard the three taps, as requested. He soon also heard taps coming from under the bed of his teenage daughter. The presence then began to communicate with him frequently, even becoming violent. His children were allegedly attacked by an invisible entity during the night. The nocturnal incidents ended with the final departure of the fearful family from the house.

The séances were, yet, a new awakening of the phenomena in the 19th century. Those that attended the séances were not just passively

14

receiving the words of the "spirits,"—they also asked questions of them about subjects from poetry to early metaphysics. Long "spiritual treaties" were written over the course of those meetings as séances became increasingly popular with the elite of Europe.

For instance, Sir Arthur Conan Doyle, the celebrated British writer known for the creation of the Sherlock Holmes series, was tremendously interested in spiritism and in the paranormal. He, too, participated in several séances, some of which found their way into his writing.

The way the séance operates makes it ideal for a party setting. After all, what could liven a party up more than calling upon spirits? These elite circles interested in the paranormal allowed mediums to become more accomplished, exclusive and prominent members of society.

SIR CONAN DOYLE

An occurrence in rural New York State with the Fox family (a father, mother, and two young daughters) gave a powerful momentum to the spiritualist movement. The Foxs had moved into a farm house in Hydesville, New York, in December 1847. Upon moving in, they were surprised by inexplicable sounds that seemed to come from within the walls, ceiling, doors, and other places. The Fox family were Methodists, a Christian denomination founded by John Wesley (whose family incidentally had also experienced poltergeist activity when Wesley was a child).

March 31, 1848 was the night that the inexplicable events in the Fox's home took a definite turn. Their 11 year old daughter, Kate, invited a ghost to rap the same amount of times she snapped her fingers. The "entity" in her presence complied with her request. It soon became clear that the phenomenon had a sentient cognitive function that could understand Kate's request, and several tests determined its validity. Even as others tried to communicate with it, the "spirit" seemed to favor Kate the most.

The communication evolved to the point that the "spirit" would indicate a letter to be used (as the alphabet was spoken aloud) to form a word. These words would eventually evolve to form sentences. Rather than cower in fear, the Foxs embraced their "ghost" to the point of learning its former corporeal identity.

The "poltergeist" claimed to be the spirit of a murdered peddler who was lost in the afterlife. He wanted purpose, and communicating with the Foxs gave him human connection. Even John Murray, the founder of the American branch of the Universalist Church, was impressed by the events at the Fox's. Coincidentally (or not), the very same day Kate Fox started to communicate with the "spirit of the peddler," another man was having a revelation of his own.

Andrew Jackson Davis was said to have been a visionary, writer, and a healer. His fame grew so large that he was named as the "Seer of Poughkeepsie." On the day young Kate established her link of communication with the spirit world, Davis had a momentous revelation: spirits could communicate with the living. That was, if the proper mediums were established.

KATE FOX

THE FOX SISTERS

JOHN MURRAY

LEVITATING TABLE

John Murray did not openly embrace the Spiritualist movement. He was, however, deeply impressed by it. The Universalist Church did not condone meddling with spirits, but still many members of the church experienced it by their own devices. As the Fox sisters became more prolific in their communication with the spirits, they started openly demonstrating their "mediumship" to great effect. In the span of just 5 years, Spiritualism was everywhere.

There were many factors that led to the favorable outcome of Spiritualism in America, one of which was that it became demystified and therefore widely acceptable. It was compatible with the so-called Christian dogma. Outside of the more orthodox faiths, it was considered completely harmless to go to a séance one night and attend church the next. It was also fully understood that mediums weren't wizards, or some sort of extraordinary being; they were simply people, just like everybody else.

Additionally, there was the massive social appeal. In the mid-19th century, séances and the paranormal were common party entertainment for the highest social circles. It was deemed luxurious and was highly sought-after. "Sitters" would gather around a table as a medium directed the séance, linking hands and illuminated by a single candle. Manifestations such as strange scents, "specters" and even the levitation of the table became "proof" of a spiritual presence.

Then there was the thrill. Communication with the dead, while not terribly uncommon during these times, played into what is now known as the "fear theory." This describes the tendency of the majority of individuals to feel enlivened and stimulated when facing an unknown that they do not perceive as an immediate danger to themselves. The medium provided a controlled environment to commune with the dead, with what they believed to be minimal risk. To provide proof of contact, mediums would communicate through knocks or raps, smells, and even in some cases, movement of furniture. Of course what many now know is that there is always, indeed, a risk.

Given the opulent nature of Spiritualism (and the medium's significant compensation) during this era, it only made sense that frauds and scam artists would come out of the woodwork. Because mediums received sizable donations for their work and became a profitable trade, charlatans devised methods to deceive well-natured and gullible people. However, the obvious phony nature of some of these imposters only proved to confirm the real séance's validity. The true mediums began using other means of communication to separate themselves.

These methods and strategies took many forms, even voluntary possession. Inevitably, mediums began to rely heavily on something known as a rapping board, one of the earliest American iterations of the Ouija. It was a simple wooden (traditionally teak) board with the alphabet carved into it. The letters were painted over with henna, a reddish-brown dye obtained from the leaves of an Old World tropical shrub with fragrant flowers. The participants would ask their question, and the medium would run their fingers over the board. When their fingers hit the correct letter, the spirit would knock or 'rap' on the table below.

Mechanical versions were soon introduced as well. They went by many names and iterations (i.e. Spiritscope, Pytho and Thought Reader) with similar functions to the rapping board and picked letters that would spell out the spirit's response to your question. These inventions, and the invention of the Planchette ("little plank") which followed, allowed the participants of a séance to communicate without a medium at all.

OLD OUIJA BOARD

OLD "LES PLANCHETTE"

While the séance was having its moment in higher social circles, it was considered blasphemous and satanic to the superstitious, elderly, and the orthodox religious. The lower economic classes cowered at the mention of spirits speaking from the far beyond. The servants in the houses of socialites cringed as their employers conjured spirits for their amusement, but the fear wasn't greater than the need to keep their families fed.

Without a doubt, these spirits captivated the séance-goers and people really did believe they were communicating with the spirits of the dead. For the most part, people who use the Ouija today still do. It leaves one to wonder: is it the truth?

CHAPTER 02

WHO OR WHAT ARE THE "SPIRITS"?

No one knows for sure what happens after we die. Those who have had Near Death Experiences (i.e. NDEs) have not undergone absolute death, but merely a relative death. Their bodily functions ceased to perform for a brief period of time, but they came back to their earthly vessel eventually. Many people who have experienced a NDE report similar experiences, and neurophysicists and psychologists have devised plausible theories for this phenomenon. But we can not say for certain what happens when our consciousness leaves our bodies for good. How can we explain death when we can not even explain life?

So, what is life? The average dictionary defines it as, "The quality that distinguishes a vital and functional being from a dead body or inanimate matter; a state of an organism characterized by capacity of metabolism, growth, reaction to stimuli and reproduction." Can this short summation adequately describe our entire existence?

We know the difference between a dog with a wagging tail and a piece of crystal; it is the difference between organic and inorganic matter. We generally think of inorganic matter as being nonliving, yet matter like crystals (though they have no organs) can "metabolize" elements of sand and incorporate its mass in a process called Melding. In doing so, crystals "grow." They are sensitive to photons that can stimulate their growth, react to stimuli, and if left to their own devices, can even repair and reproduce themselves in a process known as "mineral-reproduction." Does this mean crystals are alive? This serves only as an example of how little we know about life itself, much less the mystery of death.

BRUT QUARTZ CRYSTAL

LAPIDATED QUARTZ CRYSTAL

Some think of death as nothing but a transitional state between lives. Antoine-Laurent Lavoisier, a celebrated French chemist, taught us that nothing is created nor destroyed. It is only merely transformed.

Thanatology is considered the study of death: its effects and even the possibility of life after death. Mainstream science considers it preposterous. Fringe science sees it as yet another possibility to be considered.

The late Dr. Elisabeth Kübler-Ross was the pioneer in thanatological studies, along with Professor Ian Stevenson and Dr. Banerjee. Her first presentation was on the trinity composition of the human being: body, mind and soul. She hypothesized on both the distinction of each of these, and how they interacted together to form the human being.

According to Dr. Kübler-Ross, the body houses the physical mass that allows a person to interact with their environment. The mind is the analytical designator of a cognitive being, and houses sensorial input of interaction and memory to provide information to our body. The soul is the ultimate interactor, and decides what must be done and how to do it. Therefore, the soul truly controls the sentient cognitive lifeform.

Compare this to a living computer system: the body is the hardware that houses the system; the mind is the software that stores programs; the soul is the operator that runs the system.

Thanatology demonstrated by using precision-measuring electronic scales that the soul has weight, even when it has no mass. At the moment of death, consenting terminal patients lost between ten and twelve grams in the overall weight of their body. It was theorized that as the soul "leaves the body," the graviton (a metaparticle-wavelength of gravity) field is distorted and dispersed, creating the weight variation.

Quantum theorists on the fringe side of hyperphysics assert that the human soul is not indeed a soul at all, but a mere strand connecting our soul that resides elsewhere, beyond the realms of time and space. A single soul may have multiple variations of itself that exist simultaneously in alternate universes, all connected by multiple strands, connected to one main strand of the "true-soul." It is difficult to wrap one's brain around, but here it is depicted in graphic form:

A single true-soul is the matrix of many individual soul-fragments, each in a distinct universe of the multiverse. Each of these soul-fragments is individually connected to their true-soul by sub-tendrils. There are countless true-souls, all with their own multiversal transconnections.

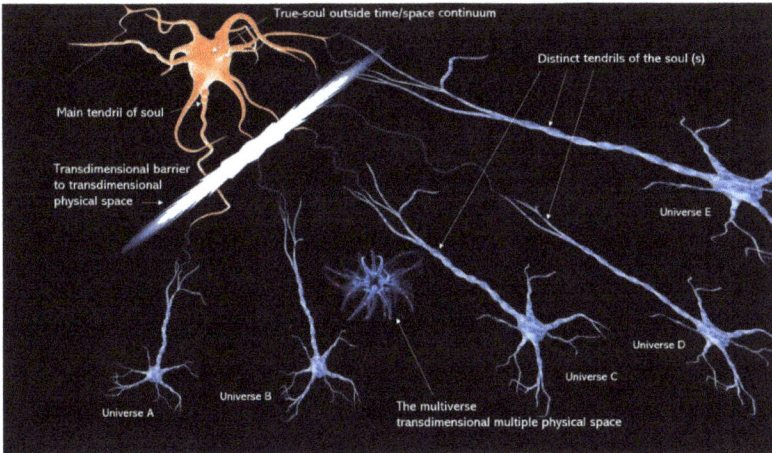

True-soul outside time/space continuum

Distinct tendrils of the soul (s)

Main tendril of soul

Transdimensional barrier to transdimensional physical space →

Universe E

Universe D

Universe C

Universe A

Universe B

The multiverse transdimensional multiple physical space

OVERVIEW OF TRUE-SOUL
AND THE TENDRILS

When someone dies, the tendril-connected soul housed within their body returns to the main tendril of their true-soul. The memory is recycled, a new "universal mode" is re-established, and the "recycled soul-fragment" is led to another life, upon another universe within the multiverse.

No soul-fragments are believed to stay in their universes after death. The "spiritual entities" communicated with during séances therefore are not the spirits of the dead we've thought them to be. They may have never been human at all.

**REPRESENTATION OF
ASCENDED SPIRIT**

Some may be an ethereal, sentient, intelligent nucleolus of life that roams the adimensional/atemporal non-space outside the multiverse's time space continuum. They are lifeforms on their own path, seeking the thrills of a physical existence and the nourishment provoked by the release of human emotions (frequencies of bioenergetic modulation are broadcasted every time we, humans, experience intense emotions).

The "tuning of the bio frequency" of our emotional broadcast can have distinctive polarities: positive (e.g. joy, grace, kindness, forgiveness) or negative (e.g. fear, hate, outrage, envy). This bioenergetic output can flow quickly or slowly, depending on our emotional intensity.

These "entities" can zero in with specific coordinates on an individual universe by using the "beacon" that is human intent. A séance is powered by human will and emotional output.

Such transdimensional entities link to the human neural system and retrieve information from them in order to know how to present themselves in a manner that will produce a more intense emotional output, namely fear. Their behavior and "memories" are not their own, but taken from the thoughts and memories of the medium and/or the sitters of the séance.

They are not the spirits of the dead, but bioenergetic intelligent predators that feed on energy broadcasted from the human participants. These are not spirits or ghosts or demons or angels, but transdimensional assailants and less-sophisticated bioenergetic lifeforms that act more on instinct than reason. Survival is a powerful motivator.

"Angels" are not as benevolent as they convey themselves to be, simply beings that strive to survive themselves by feeding on positive bioforce inputs of love, joy and other strong emotions.

The same goes for "demons" and their negative bioforce needs. Such entities can use "bioenergetic echoes" of an intense emotional event, and use it as a loophole to give the impression of a supernatural event. Hence, what we know as hauntings.

Séances are triggers and communication boards are the keys that link these entities to us. They need humans in order to do so. The entities are ethereal and cannot interact directly with our universe if not by the means of connecting to a human. It is simply energy moving energy, as matter is nothing but condensed energy on low frequencies. This seems to take the "super" from the supernatural.

So, spirits may not be the ghosts of our loved ones, but merely predators, like ourselves, seeking new prey. The Ouija gives them a tool to take advantage of their prey. But how did the modern Ouija get its start?

CHAPTER 03

OUIJA'S HISTORY

It all started because of a dream. Charles Kennard of Baltimore, Maryland read the Associated Press newspaper in 1886 that reported on a new phenomenon taking over the spiritualist community. The news article was about the talking boards and the growing demand for them among spiritualists.

Kennard had a dream about it. Something to do with a faraway Moroccan city and an object called the "Eyes of the Sky." Be it a long lost memory, chance, or the work of independent forces, Kennard seized the opportunity to make history by bringing an icon of the occult into modern existence.

His idea was to produce an improved version of the "talking board," so he changed the letter configuration, added numbers and included the words "yes," "no," and "goodbye." He also modified the board's shape to a heart-like form to symbolize life, and added a "viewing window" to focus on the chosen letter with more accuracy.

Kennard thought the inclusion of the "Eyes of the Sky" would make all the difference as the sun symbolized "life" and the moon (a half crescent with a star which is reminiscent of a symbol from Islam) linked up with his Moroccan dream. Now it needed a patent, investors, and most of all, a catchy name.

Kennard put together a small group of investors to help fundraise for a new novelty store focused around the talking board. The advice and support of Mr. Elijah Bond, a local attorney, and Col. Washington Bowie, a surveyor, helped him start Kennard Novelty Company on October 30, 1890. These gentlemen were not spiritualists, but pragmatic businessmen who recognized a good financial opportunity when they saw it.

Spiritualism had become incredibly popular at that time and Kennard's design for a talking board was far superior than any others on the market. Other boards were effective, but took way too long to produce results. This new design was as elegant as it was effective.

During a test drive at the Bond's residence with Kennard, Col. Bowie and their families, the group found that Ms. Helen Peters (Mr. Bond's sister-in-law) was a strong medium. She was able to make the talking board work and receive messages from far beyond.

The "spirit" they communed with spelled out the word "Ouija." It said that this meant "good fortune" or "good luck" in a long-lost language. Serendipitously, Ms. Peters was wearing a locket with a photo of Ouida, a famous women's rights activist. The group agreed to go with the name Ouija for their creation.

OUIJA BOARD

Mr. Kennard, Mr. Bond, and the absolutely essential Ms. Peters, went to the patent office before starting production. The chief patent officer was skeptical of the product and refused to grant a patent unless a successful demonstration was done on the spot. The creators agreed.

With Ms. Peters leading, the patent officer demanded that the board correctly spell his full name. Mr. Bond was livid, because even he did not know the name of the Polish immigrant. Kennard felt his heart sink. Yet, miraculously, Ms. Peters spelled out exactly the complete name of the officer. It was on that day, February 10 1891, that an ashen-faced and shaken chief patent officer granted Mr. Bond a patent for a new "toy or game."

The original box from Kennard Novelty Co. announced the Ouija board as "interesting and mysterious," having been proven at the patent office before it was allowed to be sold. The product flew off the shelves, even at what was then the steep price of $1.50 per set.

CHARLES KENNARD,
BROUGHT OUIJA BOARD IN USA

The Ouija board became a complete success and an absolute money maker for the businessmen. By 1892, the Kennard Novelty Co. had expanded from a single construction factory in Baltimore to other factories in New York, Chicago, and even London, England.

When Kennard died of mysterious circumstances, the company changed its name to the Ouija Novelty Company, and continued to grow in popularity. The then old and feeble Elijah Bond passed over responsibilities to Col. Bowie, who then passed the company onto William Fuld, an elevated employee and shareholder.

In 1901, Fuld became the most important man in the history of the Ouija, after Charles Kennard's legacy. He was a master of public relations gimmicks and reinvented the Ouija as an even more mysterious tool of the occult.

WILLIAM FULD

Fuld printed this on the new Ouija boxes:

Sit opposite your partner and rest your fingers lightly on the glowing planchette. Now ask your question. Concentrate very hard and watch as the answer is revealed in the message window. Will it tell you YES or NO? Will it give you a NUMBER or SPELL out the answer?

Ask any question you want. Ouija will answer.

It's only a game - isn't it?

Even though the Ouija board was now being packaged and sold as a game, it was intended to be an improved method of communication between the material and spirit world. As far as board games go, one does not generally have to fear poltergeist manifestations during Chess or demonic possession during Checkers. And certainly, if you throw a Monopoly board in your lit fireplace, it's not thought of to scream bloody murder as some have claimed the Ouija board can do. Nevertheless, if the Ouija opens doors to strange realities, one is better to play it safe than be sorry. Superstition has its roots in reality.

The Ouija kit comes with two pieces: the planchette and the board.

The planchette was trimmed into a slick heart shape with a faint glow-in-the-dark paint job, and housed a plastic viewfinder to be able to see the letters of the board through. It also had an improved three-wheel system for smooth movements.

The board became more rectangular and had a much smoother finish to allow better mobility of the planchette. The words "YES" and "NO" were printed in the top corners, with the trademarked name "OUIJA MYSTIFYING ORACLE" between them. The letters of the alphabet were printed in two rows, A-M and N-Z, and were curved into a semi-half circle. Below that were the numbers "1234567890" printed in a horizontal line, and the word "GOOD BYE" directly below it.

The corners of the Ouija board were decorated with rolling masses of dark clouds and at the top corners were the smiling sun and the crescent moon & star. Illustrations of the séance were at the two bottom corners of the board.

Fuld's kit also came with a reinvented history of the Ouija board and claimed that he had invented it. He claimed that the name "Ouija" was formed from the French "oui" and the German "ja" (both meaning "yes,") and although this explanation made little sense, it worked well for marketing purposes.

The Oxford English Dictionary defines Ouija using Fuld's explanation, but researchers of the Ouija overwhelmingly see this definition as nonsensical and childish. The truth is that we do not know all the secrets under lock and key of the elusive vaults of certain intelligence agencies, and they most likely do not know everything there is to know either. The phenomena of the Ouija still remains largely a mystery.

William Fuld was never a true believer in the Ouija board and simply thought of it as a product to be commercialized and profited from. He was a nitpicky, micromanaging workaholic who was constantly devising new ways to exploit the Ouija to make money. Little did he know that when you look into the abyss, the abyss is looking right back at you.

The following was on a footnote of the original dossier. We have no means to verify if it is true or not, but it was included in an intelligence report of the "Division Q" of the Federal Bureau of Investigation.

Entry # 203 – from the personal journalist of William Fuld circa 1926. What a crazy world we are living into! Rockets! Can you believe that? This guy, Goddard, talking about rockets. What good would that do to us? And I can't believe we will have a flying army now! Since July they said we have an Army Air Corps. What does that even mean? To me, they should... Again! Again, is this "tap-tap" noise in the walls. I tell people that I don't scare easy, but this tapping is upsetting me.

According to prior entries, Fuld complained of hearing strange sounds tapping within the walls since the previous summer of 1925, the same week he had bought the recently published novel, The Great Gatsby by F. Scott Fitzgerald. At the end of that year, he started to see phantoms and shadows moving around his home.

Fuld started to wonder how much these strange occurrences had to do with his dealings with the Ouija. Skeptically, he decided to participate in a séance. Be it either by shame or shyness, he decided to keep this event recorded only in his private journal.

It is uncertain how or exactly when this happened, but Fuld's journal relays that the séance was a private meeting with an unnamed medium and two of his most trusted employees at none other than the historical house of celebrated author Sir Edgar Allan Poe in Baltimore, Maryland. Fuld's circle of influence was strong and important enough to grant him access to this historical site after hours.

The session started at midnight and nothing of interest took place for hours except for the usual "spirit communication" that spoke of irrelevant and unverifiable events. But as the sitters were ready to break the circle and end the session, something strange happened. The clock's bell rang three times, marking the time as 3AM.

Wind blew through the house despite all the windows being closed. The temperature dropped considerably. The smell of sulphur wafted through the air and the flame of the single candle extinguished. As one of the sitters tried to relight the candle, it spontaneously erupted with a new flame, startling the participants. Then a strange, guttural and indisputably male voice came from the female medium causing dread and horror amongst them.

"Defiler! Vile Foe! You made a mockery of us! We will have your blood for that! The vengeance will be ours by right! Fatti Maschii, Parole Femine (Manly Deeds, Womanly Words)! Est Tu (It's You)!"

The medium pointed at Fuld as these horrible words left her mouth, leaving the sitters shaken.

Fuld refused to use the Ouija or participate in a séance ever again. The events remain undisclosed since, and he forbade his companions to ever speak of them. The poltergeist-like events and phantom images started one week after that night, and happened to him alone. He tried to keep constant company, but his work and responsibilities did not allow him to always have people around, especially at night. Even the dogs and cats of his neighborhood fled in terror as Fuld approached them. He lived in a constant state of anxiety.

William Fuld died the next year in 1927 while supervising workers on the roof of the main Ouija board factory in Baltimore. The iron support that held the platform on which he was standing gave way. Fuld tumbled backwards, attempting to grab the sill of an open window which suddenly closed. He then tumbled three stories, crashing on the sidewalk below and breaking several ribs. Yet, Fuld was expected to survive this accident. On the way to the hospital, a bump in the road caused a piece of his fractured rib to pierce his heart and he unfortunately perished. The workers were perplexed as to how the iron support became so twisted up and could not make sense of it. Hushed stories started to circulate among the workers of strange events happening at the factory at night, though none were ever confirmed or verified.

OLD LOGO PARKER BROTHERS INC.

Fuld's surviving relatives continued to produce a variety of talking boards until they received a business proposal to sell the company to the Parker Brothers. They happily relinquished the company, and the toy giant manufacturers revamped the Ouija by making a smaller board and creating a new plastic planchette.

Thanks to Hollywood movies and our innate thrill-seeking human nature, the Ouija is still as popular now as it was in the 1800's. But very few remember the words of advice Charles Kennard gave before he died:

(1) Do not let the board consume you. Be its master not its servant. Do not be obsessed by it.

(2) Never use the board when the user is alone. At least two people are needed to keep the "spirits" under control.

(3) Never use it around three o'clock in the late of the nights.

(4) Use the Planchette's view finder as your "spirit window," to see what human eyes cannot see, otherwise. But beware of what you could see there because It will be seeing you, as well.

(5) Never invite an unknown "spirit" in a conversation. If It doesn't give you Its name, It is evil. Be away from it. To know if It speaks Its true name, make It repeat it three times. It cannot lie three times in a row. Command It to say Its name and count how many times It will say it.

(6) Never use the Ouija on a Friday nor in a place which a death had occurred. It will be like inviting evil over.

(7) Get the name of the entity that you are dealing with. Names are powerful tools of control. Hold the name of the "spirit" and you have power over It.

(8) "Spirits" feed off your fear of them. Face the "spirit" with bravery and it will cower from your mighty will. Fear them and be cowardly and it will be your doom.

(9) Never end an Ouija session without saying Goodbye and having the "spirits" say Goodbye, as well. Otherwise you will be leaving the "spiritual door" open to whatever might come through it. To forget that is to invite evil upon you and your people.

CHAPTER 04

OUIJA: CLAIMED BY SPIRITS

Regardless that we have already established the non-existence of spirits, at least in the popular sense of the word, it remains the most prevalent explanation for how the Ouija works. Such belief has taken ground with deep roots, besides what commonsense would indicate. The prevalent claim remains that the board is used to improve the communication between us, the living, and the realm of the dead.

To these believers of a supernatural world, the astral plane (the realm of the disincarnated spirits) is a place with higher, less physical (if any at all) attributes, composed of much finer and superior material than our own plane. Such a realm of the ethereal contains:

*   The spirits of the dead from our world;
*   The spirits of the dead from other worlds;
*   The spirits that have yet to experience life;
*   The spirits that are lost and can not move into the realm of the dead before resolving their problems;
*   Wicked spirits who cause astral mischief (poltergeists);
*   Animal spirits lost within realms (animal possession);
*   Celestial guides to higher spheres;
*   Angels;
*   Demons;
*   Spirits of nature (water, fire, air, earth and lifeforce);
*   Spirits of emotion (e.g. passion, rage, euphoria, creativity, etc.);
*   Spirits of desire and lust (incubi and succubi);
*   Others

Mediums are the living bridges between these realms, and the sitters are the trigger that move the spirits. Séance and all spiritual invocations (the Ouija included, of course) serve as points of contact between realms and beacons between the planes of existence.

Astral intercommunication, according to Rosemary Ellen Guiley's Harper's Encyclopedia of Mystical & Paranormal Experience (Edison, NJ: Castle Books, 1991, page 607), is the "mind-to-mind" communication of thoughts, ideas, feelings, sensations and mental images available to the medium and the sitters. This is a, usually temporary and limited, form of possession over the mind of the medium and/or the sitters, who willingly relinquish some control to the "spirits," in order to establish communication.

One must be very careful when dealing with such entities, as they have their own agenda as the main focus of interest, regardless of the medium or sitters wishes or expectations.

Humans have their own ethics and personal moral compasses, but these "spirits" are moved by other personal and powerful interests. To most of these entities, the ends justify the means to obtain them. They are as crafty as they are cunning, and they can be very insidious in how they approach their dealings with humans.

Such beings can show a wide range of moral, intellectual and emotional qualities in order to charm, confuse or horrify their audience. The true name factor, or their true identity, is the real question. They can present themselves as the spirit of George Washington, or one's long late Aunt Petunia, all depending on their plans.

They could claim to be the spirit of the Egyptian Goddess, Isis, or a mere peddler murdered by thieves. They could claim to be demons or the innocent spirits of children. But you should not be fooled by such claims. Whoever they truly are, most likely, will not be who they claim.

By deceiving the medium and the sitters, the entities themselves (with rare exceptions) are the sole beneficiary. These entities seem to compete between themselves for a foothold onto our realm of the physical. Be them lonely or bored, or whatever their true reason may be, they masquerade themselves in any way they see fit for a point of connection between their realm and ours.

Some of these beings may not be malicious in nature, but that does not make them less dangerous. They are mercurial and volatile, exhibiting bipolar tendencies. The less innocent entities that introduce themselves with deception thrive from negative emotions, and fear can drive them into a frenzy like a shark's bloodlust.

There is always the possibility of consorting with the dangerous wolf wearing sheep's clothing, which inspires much criticism aimed at the Ouija boards themselves. Yet, the problem isn't the Ouija per se, as it is merely a tool, but the person or persons using it.

This is not to say that all "spirits" should be seen as negative and dangerous foes. Take, for example, Pearl Curran who was a housewife from St. Louis, Missouri. On the night of July 8, 1913, she allegedly received a message on her Ouija board that said, "Many moons ago I lived. Again, I come. Patience Worth, my name." This message changed the life of Mrs. Curran forever.

MRS. PEARL CURRAN

The Peculiar Tale of Pearl Curran and Patience Worth

It all seemed innocent enough: a lazy summer night of cheap thrills spent playing with the Ouija board with family and friends. Mrs. Curran was a simple middle-class housewife with a limited education, but soon found herself in a close friendship with the spirit channeled on that night. As Pearl learned more of the (alleged) formed life of Patience Worth, she opened herself up to the interesting entity.

Patience told Mrs. Curran that she was born in Dorsetshire, England in 1649. She then immigrated to the New World, seeking a fresh start among the newfounded colonies of the United States. She never got the chance to do that. She was killed in a clash with Native Americans that ended in a horrible massacre.

Pearl became fascinated with Patience's stories and began to connect with the entity directly mind-to-mind, using the Ouija board as a focus point only. She felt compelled to write about the things Patience told her, but due to the personal and unusual source of information, she did so in the form of stories.

Soon the housewife became a prolific writer. The knowledge presented to her allowed Pearl to write historical novels, poems, plays, short stories and even a monthly magazine. It was as if the minds of Pearl Curran and Patience Worth became one in the same. The well of information seemed it would never end. Six of her books were published and became great literary successes. Pearl Curran was able to not only write about Patience Worth's life experiences but collaborated with the entity to write about events prior and after Patience's alleged demise.

Literary critics did not know what to make of such a bizarre story. Nandor Fodor's Encyclopedia of Psychic Science (New York University Books, 1966, page 276) says, "Dr. Usher, professor of history in the Washington University consider [Curran's] *The Sorry Tale* a composition of 350,000 words the greatest story penned of the life and times of Christ since the gospels were finished."

These critics considered Curran's work not only well written but with such accurate historical details that she could be considered a scholar on them, uncanny for a housewife from St. Louis. They were troubled by Curran's admission of the whole Patience Worth affair. The connection between spiritualism and literature was considered preposterous, and practically forced Curran to publicly deny that the Ouija board was responsible for her abundant literary inventiveness and productivity. Her admirers and fans were baffled and refused to believe her recounting of events and the dismissal of Patience Worth; they believed she had buckled under the pressure and criticism from outsiders as she was not the only person using "automatic writing" (the phenomenon of writing collaborated on with a paranormal being, rather than the conscious intention of the writer themself).

How did a housewife with no background in writing, who had never left Missouri, become such a prolific and seasoned writer with the historical accuracy of a scholar? It is a mystery we still do not have the answer to. But if Patience Worth should be considered, it is an example that not all paranormal entities are dangerous.

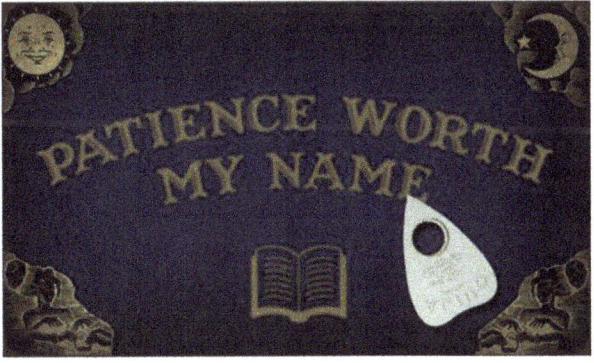

BOOK OF PATIENCE WORTH

Yet, we should not let our guard down. More so, we should scrutinize even greater the entities encountered in these cases. The entities insist on being the spirits of departed human beings. There is no way to substantiate these claims as these entities have access to a well of knowledge that provides them with all they need to know about something or someone that once existed.

Personally, I do not believe these entities to be the spirits of dead people. I cannot deny the existence of them, but I also can not confirm the true origin or nature of their existence. I understand them as adimensional and atemporal ethereal entities who have access to the knowledge, linguistic terminology and idiosyncrasy of human beings.

There are scientists that think they do not exist at all. I believe these researchers to be wrong; these entities exist, though they are not who or what they claim to be. Our scientific theories combined can not fully explain these entities and the effects they cause upon us and our environment, be it physical, social, or emotional.

CHAPTER 05

WARNINGS ON THE OUIJA

Jane Roberts was using a Ouija board when she first contacted Seth, an incorporeal intelligence residing outside the space time continuum. Like Pearl Curran, she soon started to channel Seth and went on publishing Its teachings in a series of successful books that launched the New Age movement. For 21 years, Roberts went into trances where Seth spoke through her and dictated information on subjects such as reincarnation, adjacent planes of existence, the universe, death, God and many others.

Some experts consider the Ouija to be highly addictive for some people. It could be considered the occult equivalent to an addictive drug, as hardcore devotees open themselves up to the intoxicating thrills of the Ouija sessions. Over time, it leads the way for darker forces to create a pattern of aggressive, destructive, or malignant behavior.

An account of such darkness started in October of 1933, when the then fifteen-year-old Mattie Turley decided to play with her new Ouija board for the first time. The lonesome young girl was shy and mousy, always cowering from her very strict father and condescending mother. The girl was not well considered in her social circles, as she inherited the less gracious features of her father instead of the natural beauty of her mother. Known to the other girls as "Fatty Mattie" and "Woebegone Mattie," the young Turley was prone to do things by herself.

While her parents were having dinner with friends elsewhere,

Mattie contacted the "spirit" of Magda, an old Gypsy woman that had died a long time ago. Magda told Mattie that she was sorry for her and the awful things in her life. Magda brought along another "spirit" named Yacobbo, a "seer" that could help Mattie be happy in life. The "spirits" told Mattie that her mother was being controlled by her father, who was having romantic affairs with other married women. Mattie was scandalized by this knowledge, and became indignant towards her father.

The "spirits" told Mattie that if her mother met a good man, a fun man, a hard-working salt-of-the-earth type of man, she would change as a person and they would become closer than ever before. But Mattie had to change her ways, quickly and drastically.

The beginning of this story starts with Dorothea Turley, Mattie's mother, who was a beauty contest winner from New York, though she was born in England. She was as beautiful as she was cunning and greedy. Dorothea seduced and married Ernest J. Turley, a retired Naval Officer from a good family with solid finances. She gave birth to Mattie, but wasn't happy as she thought she had "lost her figure" with the pregnancy. To make matters worse, Mattie was an ugly baby in her opinion who looked more like her father and less like the daughter of a beauty contest winner.

Mattie grew up with a distant mother and a workaholic father, who instead of having affairs with married women, worked hard on his business to provide for his family. The supperless nights were due to working overtime, not sinful trysts with adulterous women. But Mattie did not know this.

As the years passed and Mattie became a teenager, her mother became sick and tired of the metropolitan life in big cities. Dorothea convinced Ernest (who was quite in love with her and unaware of the loathing she felt for him) to move to Arizona to fare better with her health. To new beginnings and better chances, the family packed up and moved.

Once there, Dorothea became involved with new friends who introduced her to the Ouija. She became fascinated with the talking board. During one of their sessions, they contacted the "spirit" of the old Gypsy named Magda. She became close to the entity and soon began to contact it by herself. She said Magda understood her feelings like no other before; that the "spirit" understood her sadness, frustration and sorrow of being married to a man she didn't love and being the mother of a child she didn't care for.

Dorothea felt that she needed to stay with Ernest because of his financial support and status. It was a harsh and difficult thing to pretend to be a lovely wife when she despised the man. The "oaf" she called him, secretly. Her daughter was also a disappointment to her, as she saw Mattie as awkward as her husband was. Magda advised Dorothea to live her life and find happiness elsewhere. In Arizona, there were many eligible bachelors that were attractive to Dorothea.

Dorothea met a young, blonde cowboy, twelve years her junior. She became infatuated with him, and soon the relationship moved beyond a few trysts. Wanting Ernest out of her life, she asked Magda what to do. The "spirit" had to call upon the aid of another, that called itself Yacobbo. Yacobbo was old and cunning, and said that in order to achieve what Dorothea desired, she had to bring the "young one" (Mattie) into their schemes.

The "spirits" forbade Dorothea from participating in the sessions alongside Mattie, and she was told to leave until their affairs were finished. Mrs. Turley thought they would convince Mattie to make her husband leave the marriage, keeping her good name. She would keep the estate and could send Mattie to a private school overseas. Confident of her plan, she went shopping to buy a brand new Ouija board to offer Mattie as a surprise gift.

During her first session, Mattie contacted the "spirit" of the Gypsy, Magda. The "spirit" quickly gained Mattie's absolute trust, and helped her get even with her bullies. Magda soon introduced her to Yacobbo who became an even stronger influence over Mattie.

Mattie became fierce and strong, becoming more confident with the help of the "spirits" she was communicating with.

Ernest, Mattie's father, was a strict but fair man. He worked hard for his family and indulged Dorothea's whims as much as he could. He considered himself a man of great fortune to be married to a beauty like Dorothea and found her to be a good wife, if not a bit eccentric. He himself did not believe in the "nonsense" of spiritualism, being too pragmatic and logical. Yet if his lovely wife liked it, he was happy to indulge her. Ernest could have been unfaithful to Dorothea without her being the wiser, but indeed he had only eyes for her. Little did he know what his beloved wife truly thought about him.

By November 1933, Mattie was exclusively contacting Yacobbo through the Ouija. She was cautious and caring, and her new "spirit" best friend always knew what to say to the girl to keep their connection.

Dorothea used the board very rarely by this time, as she was enamored with her rustic, young lover. But on a rainy afternoon when she could not meet with him, Dorothea decided to use the Ouija. She contacted Magda and was surprised to find the spirit so unsettled. It spoke of doom and how unwise it had been to allow Yacobbo to take control of Mattie.

Dorothea responded that, to some extent, it was a good thing because the girl was finally starting to grow a backbone.

Magda warned Dorothea that she was responsible for sealing her own fate; that Mattie could either be her joy or her undoing depending on what actions she took next.

Thinking this to be nonsense, Dorothea asserted that her daughter was just fine.

Magda called Dorothea a Jezebel, and her Ouija board, allegedly, broke in half. Dorothea never used the Ouija again, shaken with the outcome. But soon she forgot the whole episode, her thoughts turned towards her affair.

Meanwhile, Yacobbo was spinning tales to a disgusted Mattie of all of her father's (false) affairs, and how evil and controlling he was. Yacobbo told her that Ernest despised his wife and daughter, but that she would devise a plan to solve all Mattie's problems and it should all be over soon enough.

On the night of November 15, 1933, Yacobbo revealed her plan. Mattie was to convince her father to bring her along on a hunting trip over the fields of their ranch near Springville and kill him there. She was, under no circumstances, to tell her mother any of their plans, and her story would be that she had shot her father by accident while trying to hit a skunk. Her mother would then be a widow and capable of marrying a certain handsome cowboy who would become her new dad. Mattie's evil father would be dead, her mother would love her, and she would have a much better new father if only she had the courage to make it happen. Naively, Mattie believed the "spirit."

On November 18, 1933, fifteen-year-old Mattie Turley shot and critically wounded her father using both barrels of a twelve-gauge shotgun at their ranch. It almost didn't happen. In the moments before the shooting, Mattie swore she heard an old woman's voice whispering in her ear, "Don't do it, girlie. It ain't right." As she was considering it, frozen, her hands seemed to move by themselves. She took aim and shot her father in the back, surprised as she had no recollection of pulling the trigger herself.

The gunshot brought people quickly to the scene and soon help came along. Ernest was badly wounded and unconscious. Ernest died from these wounds six weeks later, and the police decided to investigate the "sad accidental death" of Mr. Turley. Contradictions were soon found in the story as Mattie and Dorothea retracted their depositions.

Mattie finally broke down and told the police that she had shot her father on purpose, but it was the fault of the spirits she had consorted with while using the Ouija board. She claimed that the "spirits told [her] to kill daddy so that mother could marry the cowboy."

The police didn't believe her story, and assumed that Dorothea Turley had plotted to kill her husband by having her daughter shoot him, thinking she wouldn't be punished as harshly for the deed.

The defense argued that the shooting was an accident, and Mattie had indeed been aiming at a skunk when she hit poor Mr. Turley. They insisted that the police had forced Mattie to confess to a crime that she had not committed. Dorothea asserted that she "knew nothing of the Ouija board's instructions to her daughter and...the girl was talking insanely."

Regardless of the efforts of the defense, Mrs. Turley was convicted and sentenced to ten to twenty-five years in the State Prison at Florence Institution. A later appeal and three years of deliberations later, Dorothea's sentence was overturned.

As for Mattie, she received six years in the State Reform School for girls in Randolph, where according to certain accounts, "the little girl told the reform school matron that she was sorry she killed her father but she was certain of the Ouija board's instructions."

The Turley story may serve as a cautionary tale of the dangers of becoming obsessed with the Ouija board. An entire family was destroyed by either their own personal demons, or had them awoken by entities of the far beyond.

THE TURLEY FAMILY CASE

Of course, not all people who use the Ouija board will end up rotting in a prison cell, dead of mysterious causes, or tied to a bed like a demonically-possessed person read the "Rituale Romanum" by an exorcist. Yet, they certainly could. To wander into the unknown is to invite disaster if you do not know what you are really doing.

Not all entities are as wicked as Yacobbo. Magda certainly seems to have tried to avoid such disgrace. And yet, neither were indeed who they said they were. True benevolence doesn't come with deception: both entities had motives to wreak mischievous havoc on the Turleys, and indeed they both fed well from the outcome of such an emotional maelstrom of tragedy.

A rumor began to spread amongst the few that knew of Mrs. Turley's affair with the young drifter from Texas that was working as a cowboy on a nearby ranch in Springville, Arizona. It was said that the body of the twenty-three year old named Kent Pearce had been found dead, the victim of several knife wounds to his back. Besides Mrs. Turley, he also had a non-identified teenage male lover referred to as "Dan."

Dan was allegedly fifteen or sixteen, the son of a local farmhand. It seemed that Kent had taken a fancy with the teenager and seduced him as his second secret lover. Dan was not aware of the affair between Kent and Dorothea, or the monetary gratuities she provided him with.

Kent felt he had the best of both worlds. Dorothea had spoken to him about the Ouija board many times, so Kent decided to surprise Dan with one as a gift. He thought that it would amuse the boy, making him even more captivated by him. But in doing this, he guaranteed his own death.

The first time Kent and Dan tried the Ouija board together, a "guardian angel" named Y presented itself to them. Kent didn't believe it, thinking it was Dan moving the planchette. After this, Kent had no interest in using the Ouija board again, so Dan started using it alone. Y ingratiated itself with Dan who soon became close to his "angel."

Y was wicked in its own ways. It convinced Dan that Kent didn't love him at all, and that he had a secret lover: an older but very beautiful woman from a nearby ranch. Y told Dan that the woman wanted to marry Kent as soon as she became a widow (somehow knowing that her husband was due to die soon) and that they would laugh at the "silly green-eyed boy" that thought he had won the heart of the handsome cowboy. Dan became furious at Y for "saying such lies," but Y dared Dan to find out himself.

Sadly, Dan verified that Y had not been lying. From a safe distance, he spied on Kent with Mrs. Turley. Furious and heartbroken, Dan swore revenge on Kent. That same night, Y advised Kent on how to get revenge and run away from his house, town and sorrowful life.

Dan brought Kent far into the woods and they made love next to a campfire. Afterwards, while Kent rested peacefully, Dan retrieved a knife from its hidden place, straddled Kent and stabbed him over and over again. The young cowboy didn't even have the chance to defend himself. Dan watched, detached, as his lover bled to death, blood smeared all over their naked bodies.

When Dan realized what he had done, he vomited until there was nothing left in his stomach. He began to cry for Kent, for himself, and for having listened to the wicked Y, who certainly was no guardian angel. The boy kissed his late lover one last time, then ran away that very night and was never seen again.

So, how did we learn what became of Kent and Dan?

That was where the rumor started: from a "spirit" on a Ouijia board. According to the rumor, the "spirit" was said to have committed suicide. The name of the spirit was said to be Dan.

The Ouija board is a well of lore and strange tales, as you may be coming to find. Edgar Cayce, the famous "Sleeping Prophet" of Kentucky, warned about the many dangers of the so-called "automated writing." Cayce referred to the Ouija board as a "dangerous toy," a sentiment that many other mystics shared. For instance, "Psych News," a noted English spiritualistic publication, began a popular campaign in 1968 demanding the ban on the sales of Ouija boards. But the strongest attacks came from Christians who cited the Bible's prohibitions against divination and consorting with the dead.

Even today, websites such as The Lifehouse and Evangelical

Outreach warn that involving oneself with the occult could lead to witchcraft, Satanism, demonic possession and eternal damnation in the lake of fire.

Then there is the controversial case that appeared in the Washington Post newspaper on August 20, 1949, titled, "Priest Frees Mt. Rainier Boy Reported Held in Devil's Grip." The story was about a boy that, allegedly, became possessed by demons after playing alone with a Ouija board.

In a deeper investigation, substantiated by the work of Mark Opsasnick's "The Haunted Boy of Cottage City: The Cold Hard Facts Behind the Story That Inspired 'The Exorcist'," the accounts by the Washington Post were poorly documented and unreliable.

However, the point stands: Ouija boards seem to be dangerous in the hands of unqualified or inexperienced "sitters." But are ethereal entities indeed behind these alleged paranormal events?

Mainstream science thinks otherwise. Let's explore their point of view in the next chapter.

CHAPTER 06

THE OUIJA AS SEEN BY MAINSTREAM SCIENCE

Many scientists say that Ouija boards are <u>not</u> powered by spirits or demons. And they are absolutely right, even if for the wrong reasons.

Experts affirm that there are two basic principles that explain the Ouija effect. One is called Automatism, known more commonly as the Ideomotor Effect. The other is named Disassociation, a psychological condition.

To these professional authorities, the Ouija effect happens as a consequence of these two principles working in unison.

In 1852, William Benjamin Carpenter, a physician and psychologist, published a treaty on automatic muscular movements that happen without the conscious will or volition of an individual. For instance, think about the emotional response you may have upon viewing a sad scene in a play: the glands of your eye ducts begin to discharge tears. Or more simply, even the basic action of breathing happens automatically. As other researchers read Carpenter's treaty, they drew a possible link of the Ideomotor Effect (Automatism) to the events reported during séances.

The next year, Michael Faraday, a chemist and physicist, conducted several experiments that demonstrated the possibility that the table's motion happened due to the Ideomotor Effect. Faraday's experiments were not conclusive, but nevertheless were seen by experts of anomalistic psychology as strong and factual. They concluded that Automatism caused involuntary muscle movements without the

conscious knowledge of the participants at séances. An adequate theory, the Ideomotor Effect could certainly explain the events of some of these paranormal stories.

But does it explain all of them? Although many experts are vehemently vocal against the study of psychometry, perhaps we should not be so quick to discount it.

Psychometry is defined as the ability to sense and locate distinct frequencies at a site, to measure approximately its volume and/or quantity, and to provide a location for the searched compound such as:

1) Water - as a psychometrist seeks water resources
2) Air - as a psychometrist seeks wells of gases or gaseous components
3) Earth - as a psychometrist seeks metallic-based elements (i.e. gold, silver, iron, etc.)
4) Fire - as a psychometrist seeks mineral-based elements (i.e. diamonds, crystals, etc.)
5) Lifeforce - as a psychometrist seeks a connection or retrieval of the frequencies of individual biosignatures (e.g. the location/impressions/data of missing persons)

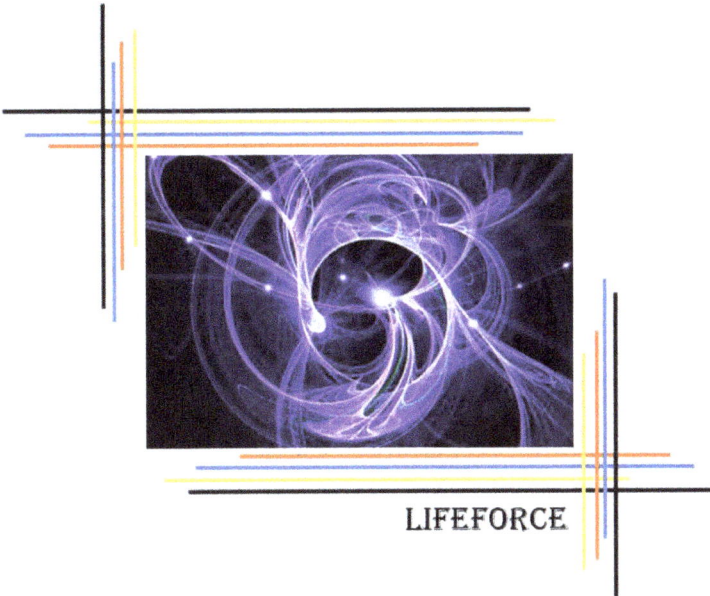

LIFEFORCE

Although already stated that each individual human has their own unique and specific biosignature (or the equivalent of a broadcasted fingerprint), mainstream science has yet to recognize their existence, much less the existence of psychometry.

Because most psychometrists use physical devices as a grounding "psychic-anchor" (i.e. dowsing rods, pendulums, talking boards, planchettes, etc.) that pinpoint and regulate what they are searching for, experts insist that they work based on the same principle of nonconscious movement. Since many of the devices used require diminutive muscular movements to work and are made of lightweight materials, scientists assert that psychometrists subconsciously move the tools themselves.

The flaw in this theory is that if one becomes conscious of the involuntary movement, then it becomes voluntary. So what, then, moves the planchette if not involuntary movements, especially in the rare reported cases of planchettes moving on their own without human contact?

We must also take into account the theory of Disassociation: the person allegedly moving the planchette across the Ouija board has no consciousness of such action, their subconscious mind controlling their actions. In this condition, the "affected" would be temporarily disconnected from their own consciousness, allowing a distinctive aspect of his or her personality to take control of the action. Scientists substantiate this claim by pointing to the existence of Dissociative Identity Disorder (previously known as Multiple Personality Disorder), whereas an alternate personality would suppress the main consciousness and take control of a person's motor skills.

Nevertheless, this explanation does not explain the knowledge the medium or "sitter" accesses that neither the conscious or the subconscious mind had previously known. To point to psychology is indeed valid, but to ultimately present it as the solution for the Ouija's workings is the same as to say that even when a house is made of stones, a pile of stones is the same as a house. None deny the ability of the subconscious or unconscious collective, but to assert that all Ouija-related episodes are a consequence of them is both preposterous and irresponsible.

A more sensible conclusion: some of the Ouija's workings happen due to a combination of Automatism and Disassociation (one a neuromuscular aftereffect and the other a psychological condition). And some are provoked by an external agent (or agents): not the spirits of the dead, but a sentient cognitive entity of ethereal consistence (something like a bioenergy that is coherent and aware of both itself and its environment, either intelligent or merely intuitive, and assumes the role of a predator seeking prey).

Referred to by some as "miasmas" (the dictionary refers to it as a form of swamp gas caused by decomposition), paranormal "entities" are made of coherent bioenergy that, allegedly, exist between the distinct frequency fields of harmonics. These harmonics separate the unique modulations that compound the single universes of the multiverse, including our own universe.

Such miasmas would live at atemporal and adimensional non-space, outside of our own multidimensional time space continuum. Since no proven evidence of this can be provided, it remains in the realm of theoretical metaphysics. Although mainstream science is still reluctant

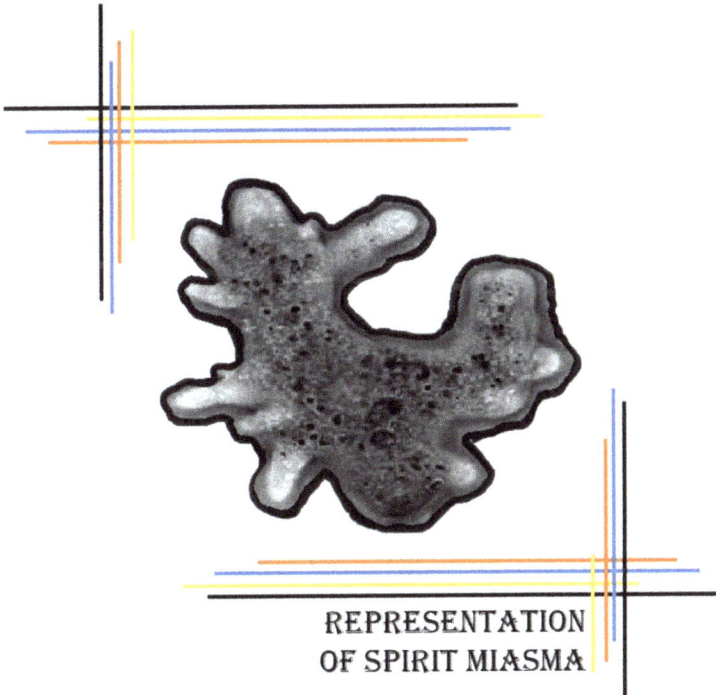

REPRESENTATION
OF SPIRIT MIASMA

to accept its existence, there are a few innovators and advanced theorists exploring the possibility. We must take into account that if not for these daring and brave scientists, who face ostracization and becoming the laughing stock of the science community, we would still be learning that the world is flat and that Earth is at the center of the universe.

Science, in order to progress, must continually reevaluate and explore outside of its comfort zone. Our "basic laws of physics," everything we know and take for granted, should not be considered as conclusive, immutable facts. As knowledge changes, so does our understanding of it.

In Sir Isaac Newton's Laws of Motion:

1) An object in motion moves at a constant velocity in a straight line unless acted upon by force. Likewise, an object at rest will stay at rest - a property known as inertia. [Telekenesis states that a rested body can be put in motion by gravitic force once the gravitation (theoretical metaparticles of gravity) field of an object is restructured by a change of modulation of its integrity field annulling its inertial stance moving a stationary mass.]

2) The acceleration of an object is proportional to the force acting upon it and inversely proportional to the mass of an object.

3) Force (F) equals mass (M) times acceleration (A): F=MA [Telekenetic impulse can alternative the velocity of tractive motion of a given mass independently of its quantity of matter by increasing or decreasing the flow of gravitons directed upon it.]

In Sir Isaac Netwon's Law of Gravity:

In common usage, gravity refers to the force between planets/objects on or near them. In scientific parlance, gravitation represents one of four basic forces controlling the interactions of matter. The others are strong and weak nuclear and electromagnetic forces. Some consider also including photonic forces, but since a photon is a quantum electromagnetic form of radiation (i.e. a wave particle), it is already included within the vast spectrum of electromagnetic powers.

The gravitational force (F) between objects is proportional to the products of their masses (M1 and M2) and inversely proportional to the square of the distance (D) between them. "G" represents the gravitational constant in Newton's Law of Gravity, a fixed ratio of approximately 6.67390 x 10^{-11} newton M^2/Kg^2, (that should be taken in consideration the action of external variable forces that would alternate the gravitational constant making the fixed ratio to be... well... not fixed, but indeed in a state of possible fluidity making such fixed ratio unreliable, under certain exceptional circumstances).

The basic law of gravity is:

$$F = G\frac{M^1 M^2}{d^2}$$

Earth's gravitational force is considered to pull objects toward it at a constant acceleration of g=9.8m/s^2. This allows calculations of the velocity (V) of an object with an initial velocity of V^o (since it was under inertia [a property of matter] whereby it remains at rest or continues in uniform motion, unless acted upon by some outside force) in free fall at a given point in time (T), and calculation of the distance (D) of an object from earth at any given time, with any given initial velocity (VO) and any known initial height (A) via the following equations:

$$V = V^o - gt$$

$$D = 1/2g(t^2) + V^o\, t + a$$

Assuming that height is measured in feet and speed in feet per second, the maximum height (h) reached by an object with positive V^o is expressed as: H

Several aspects of earth slightly distort its gravitational force. Gravity is lessened by the centrifugal effect of the earth's rotation. At the poles, where centrifugal force is absent, acceleration of gravity is greater.

Also, the further an object is from earth's center, the smaller the gravitational force. Gravity is weaker upon mountaintops than at sea level.

And, it is the external factor to be considered here, the

antigravity effect caused by an outside force that could isolate the mass of a given thing from the gravitational force by surrounding it with a considerable neutrino field (a neutrino in an uncharged elementary particle held to be massless or extreme light to the point of almost lightness) that would isolate it from the influence of gravity. Again, it is the principle of telekinesis workings, a mean that could be used and/or exploited by the theoretical miasmas.

Conservation Laws

In physics, laws of conservation state that in a closed system, where neither mass nor energy is added or subtracted, certain measurable quantities remain constant. At a sole unidirectional dimensional point-of-view, that is correct. But, when there are overlappings of distinct dimensional harmonics structures on a given point, conservation is annulled by external manipulating forces. A transdimensional conduit (also known as a quantum anomaly) is formed and a vortex pathway is opened between dimension to dimension, or admimensional atemporal non space to dimensional temporal space.

*Conservation of mass: Mass is neither created nor destroyed within a closed system except when converted to other basic energies that could occur by dispersion, conversion, diversion or convergence.

The alleged miasmas are said to master the ability of conservational multiconfigurations producing energy manipulation.

All moving objects have momentum, and in a closed system, total momentum is always conserved. Linear momentum is the product of the mass of an object and its velocity. In the following equation, "M" and "V" represent the initial total mass of an object and its velocity within a closed system. After a collision between those objects, the mass and velocity of individual objects may change (for example, one object could break into smaller pieces, each traveling at a different velocity) but the product of the total mass and velocity in the system after the collision (MV) will remain the same.

But, any object moving in a circle has another kind of momentum: angular momentum. This is because circular motion requires acceleration toward the center of the circle. The amount of acceleration

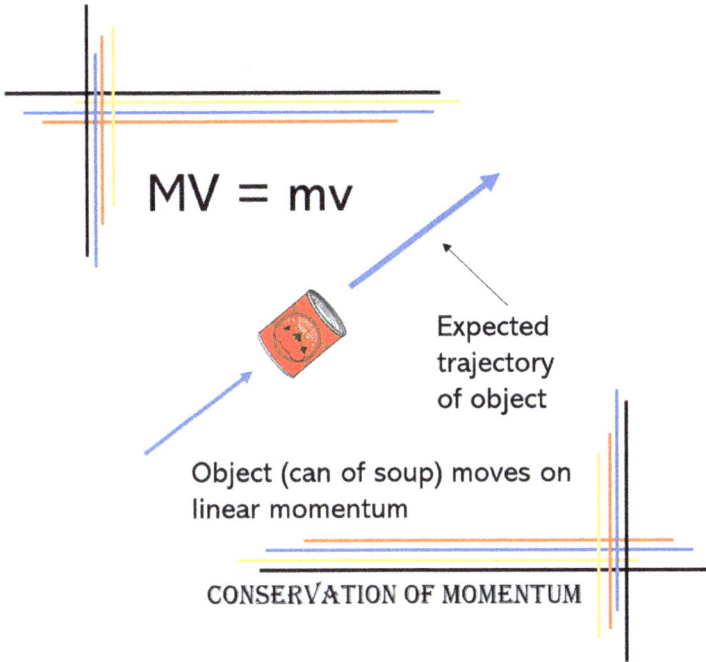

MV = mv

Expected
trajectory
of object

Object (can of soup) moves on
linear momentum

CONSERVATION OF MOMENTUM

depends on the speed of the object and the square radius of the circle
(angular momentum is the product of this speed of the mass of the object
and the square of the radius.)

$$Ag = \text{Angular Momentum}$$
$$M = \text{Mass}$$
$$\text{Radius}$$
$$Ag - V + M + R^2$$

The total amount of energy in a closed system will not change
except when converted to mass. Or in some cases the equivalent of mass.
That is how kinetic force is achieved by the miasmas influx of external
gravitic force, creating a gravitational pull equivalent to the mass when
there is no mass provoking an action.

Albert Einstein's special theory of relativity states that mass and
energy are related. Because one can be converted into another, mass and
energy <u>alone cannot</u> be conserved. But the total amount of mass and
energy together must be conserved.

This is reflected in the following equation where "M" is mass, "E" is energy, and "C" is the speed of light in a vacuum (which usually is constant):

E=MC²

Relativistic mass can describe how mass increases with velocity. The following equation – where "M" is the mass of a moving object, "Mº" is the object's mass when not moving, "V" is velocity in relation to a stationary observer, and "C" is the speed of light – shows the relationship:

$$m = \frac{m^\circ}{\sqrt{1-\frac{v^2}{c^2}}}$$

The theory that no object can travel faster than the speed of light is based on this equation. As an object approaches "C," so much energy is converted to mass that it no longer accelerates.

So, the key point here is – the <u>speed of light.</u>

The speed of light was first measured in a laboratory experiment by the French physicist Armand-Hippolyte-Louis Fizeau (1819-96).

Today the speed of light is known very precisely as 299,792,458km per second (or 186,282 miles per second) in a vacuum medium. In water, the speed of light is about 25% less, and in glass, 33% less. Meaning: there are variations and no constancy depending upon the medium on which the photons are traveling.

But that is the point: photons are a linear wave particle.

Now bring into consideration another model, this time a metaparticle known as Tachyon: a hypothetical particle that travels faster than the speed of light. Mainstream science posits it as theoretical, as it is not consistent with our currently known laws of physics. Yet the military (secretly, their labs) presents it as fact. Using string theory (the unification of quantum mechanics and general relativity), we can posit that the basic constituents of matter can best be understood not as point objects (such as photons) but as tiny closed loops, or "string."

Meaning, since all is connected, there is <u>no</u> linear travel. A Tachyon arrives wherever it aims to be (regardless of how far it is from it) at a fraction of the moment on which it departed so.

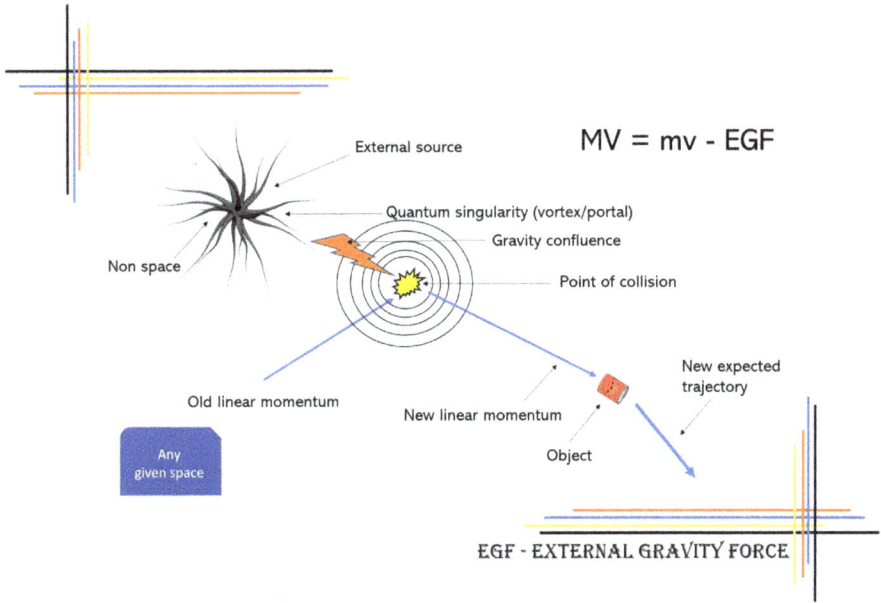

$$MV = mv - EGF$$

External source
Quantum singularity (vortex/portal)
Gravity confluence
Point of collision
Non space
New expected trajectory
Old linear momentum
New linear momentum
Object
Any given space

EGF - EXTERNAL GRAVITY FORCE

Why are there cold spots in "haunted" areas? What do these cold sites have to do with Hyperphysics?

(1) Heat is a form of energy. Within a closed system, energy must be conserved except in nuclear reactions or other <u>extreme conditions</u>. It is neither created nor destroyed.
(2) Within a self-sustaining system, heat can never go from an area of low temperature to an area of high temperature, for this would require added energy. Without added energy, disorder, or entropy, can only increase.
(3) Absolute zero cannot be attained by any procedure in a finite number of steps. Although it can be approached asymptomatically, it cannot be reached.

Why do electronic devices malfunction during paranormal events? Why are electrical anomalies and surges present during paranormal activity?

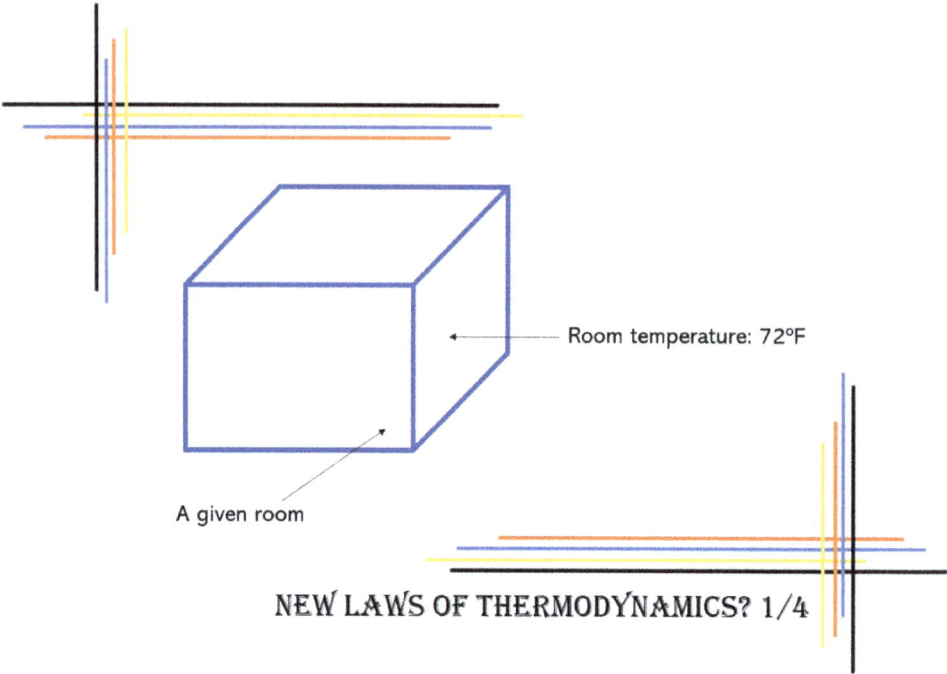

Room temperature: 72°F

A given room

NEW LAWS OF THERMODYNAMICS? 1/4

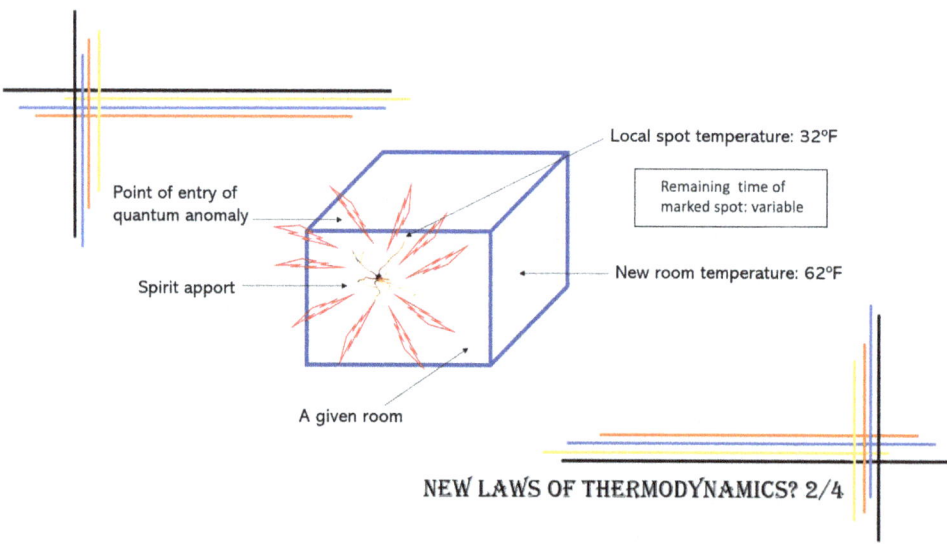

Local spot temperature: 32°F

Remaining time of marked spot: variable

Point of entry of quantum anomaly

Spirit apport

New room temperature: 62°F

A given room

NEW LAWS OF THERMODYNAMICS? 2/4

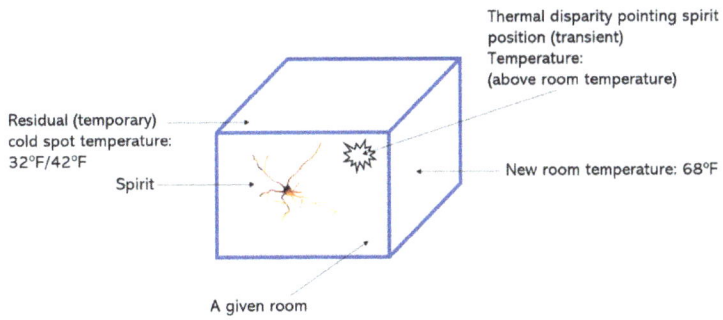

Thermal disparity pointing spirit position (transient)
Temperature:
(above room temperature)

Residual (temporary) cold spot temperature: 32°F/42°F

Spirit

New room temperature: 68°F

A given room

NEW LAWS OF THERMODYNAMICS? 3/4

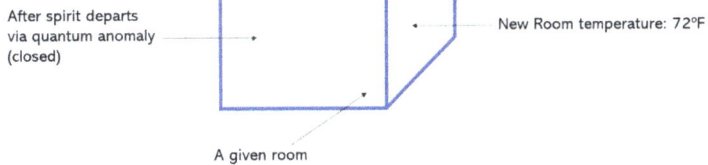

After spirit departs via quantum anomaly (closed)

New Room temperature: 72°F

A given room

NEW LAWS OF THERMODYNAMICS? 4/4

Laws of Current Electricity

Electric current generally represents the flow of electrons through a conductor. The rate at which electrons flow can be measured in amperes, defined as a number of electrons (measured in a unit called coulomb, equal to about 6,24 quintillion or $6,24 \times 10^{18}$ electrons) moving past a particular point every second. One ampere is equal to one coulomb of charge passing each second. Like water, electrons tend to move from areas of high pressure to low pressure. The difference between these two "pressures" is known as potential difference and is measured in volts.

Certain substances, such as copper and carbon, allow electric currents to pass more readily than others. That is, they have greater conductivity. Human beings, as almost all other living organisms on earth, are of course made of carbon.

We are indeed electrical-prone lifeforms. We are not surrounded by electricity, but we are basically made of bioelectricity.

The resistance to conductivity is measured in ohms.

Ohms law's electric current is directly proportional to the potential difference and inversely proportional to the total resistance of the circuit. "I" is electric current (measured in amperes), "V" is the potential difference (measured in volts), and "R" is resistance (measured in ohms) making the equation:

Electrical power "P," measured in watts, represents the rate at which electricity is converted into some other form of energy (such as light, as the photons that brighten a lightbulb). "P" is the product of current and potential difference using the following equation:

$$P = IV$$

So, what does this have to do with paranormal activity? Let me explain.

For instance, take the usual flow of electrons in a given area. The site should be stable and its flow of electrons nominal. But if a quantum singularity irrupts into this given area, things will change rapidly in consequence of this.

The electrons will be moved by displacement as the quantum singularity creates a "gravity well" as it "opens" into our universe, moving from an area of high pressure into an area of low pressure. Since it happens within a resistance of conductivity, it will create "gaps" into electric cycles of propagation. This provokes short circuits, surges, or power outages due to the increasing flow of electrons displaced by the "gravity well," that was caused as a side-effect of the opening of the transdimensional conduit (a.k.a. quantum singularity.) Hence, why during paranormal activity, there are all the usual electrical and electronic malfunctions. Electrons would flow out of sync.

The simple presence of a transdimensional entity would be a focus for anomalies of an electron's flow - the entity would be invisible to our eyesight, but electronic equipment would monitor its effects on the site. The circuit components of electrical devices would malfunction, or, at least, be affected somehow by the increase of the flow of electrons. This is due to the displacement between higher and lower zones of pressure provoked by the movement of the "miasma" (or any other entity of transdimensional origin).

So, quantum singularities, the multiverse, transdimensional gateways, bioenergetics, cognitive ethereal life forms, gravity wells... seems more like the realm of science fiction, no?

Yet hyperphysics and quantum mechanics are certainly part of science, and perhaps we should not be so quick to discount what we do not yet understand.

For instance, virtual particles are defined as subatomic particles that rapidly pop into and out of existence. They can exert real forces, usually occurring in particle-antiparticle pairs and are rapidly annihilated. Well, by following the very first law of conservation (the conservation of mass), we know that mass is neither created nor destroyed within a closed system except when converted into energy. Meaning, it is transformed.

From where did these rapid subatomic particles "pop" into existence in our universe? And, after that, to where do they "pop" out

65

after their alleged annihilation? Even if their energy was converted, they would leave residual traces behind, no? Why are there no traces of it? If something <u>outside</u> our universe enters it, logic dictates that it came from elsewhere, such as another universe or universes. Hence, the acceptable existence of a multiverse. The entry of those foreign sub-particles would imply the existence of conduits between our realms – conduits of entrance and exit. And the consequential behavior of certain clusters of sub-particles would indicate them to be part of a whole that could be considered as a lifeform if it behaves in a sentient and/or cognitive manner.

The realm of quantum physics brings even more to enlighten us about these new views that defy conventional physics. Let's start with the <u>two basic laws of quantum physics</u>:

(1) <u>Heisenberg's uncertainty principle</u>: Certain pairs of observable quantities (like energy and time, or position and momentum) cannot be measured with complete accuracy simultaneously because of the many variables that would have to be considered within it. This edict is also known as the indeterminacy principle.
(2) <u>Pauli's exclusion principle</u>: Two electrons in an atom cannot simultaneously occupy the same quantum energy state. This has since been shown to be true for many subatomic particles.

Suddenly, our supernatural becomes quite natural as we begin to understand what baffled us before. Of course, there are still a lot of unanswered questions, but we have come a long way from the dark nights we spent cowering before spectres. Nowadays, we walk into well-illuminated quantum physics laboratories.

There is an old Latin saying: "omnia mutantur, nos et mutamur in illis." It means, "all things change, and we change with them."

Spirits are not what people believe they are. Life can come in all shapes, forms and densities. Even ethereal ones. They are not the spirits of dead people but energy-based sentient and/or cognitive life forms preying upon simple people, feeding from bioenergy output broadcasted as consequence of their extreme emotional state of mind. Fear is indeed a powerful motivator.

66

REPRESENTATION OF SPIRIT

But, still, there is yet another alternative we must consider in order to be fair, regarding the possibilities on the workings of the Ouija board and its alleged spiritual interaction: <u>Constructed Ghosts</u>.

What are those? Well, we shall see that in the next chapter called <u>The Canadian TSPR and the Philip Effect.</u>

What if the "spirits," in the sense of paranormal manifestations, and the "sitters," participating in the spiritualistic encounter, would <u>not</u> be <u>two</u> distinct groups, but part of the very <u>same phenomenon</u>? What if the phenomenon originated from the participants themselves? See it in the next chapter.

CHAPTER 07

THE CANADIAN TSPR AND THE PHILIP EFFECT

In 1972, the Toronto Society for Psychical Research (TSPR) performed a most original and innovative experiment that attempted to answer three significant questions:

1. Could the séance phenomena be created in fully lit (or daylit) rooms instead of in darkened rooms (or at night)?

2. Is the "spirit" phenomena produced by disembodied spirits or by living people?

3. Is a medium or focus point (like the Ouija) necessary for phenomena to occur?

In order to create such an experiment, these requirements as outlined in the TSPR work, "Philip: The Imaginary Ghost," had to be met:

• None of the eight participants should be considered a psychic or have shown any evidence of being a medium

• The participants must create a "ghost" of their own, made up of the distinct input of all the participants

- They could use real places and historical events, but the "ghost" they conjured had to be an imaginary person

- A particular room had to be designated in order to be "quarantined" from outdoor access

- No one could summon the "spirit" alone, and the whole group of eight had to be together in order for the entity to manifest itself

This is how the Philip experiment began: The idea was to create an imaginary character that the participants would summon during a séance. The sitters were to come up with a historical, yet fictional, past and create a tragic and romantic backstory to match the loss associated with traditional ghosts.

Dr. A.R.G. Owen was the director of the project, and in charge of the eight volunteers known as the Owen Group. The group chose Diddington Manor as the location for their character they named Philip. They also created two supporting secondary characters named Dorothea (his wife), and Margo (his lover). Philip was to be an aristocratic Englishman living in the mid-1600s during the time of Oliver Cromwell. They decided Philip would be young, reputable, Catholic, and a supporter of the King. His wife, Dorothea, would be young and beautiful, but also frigid and distant. She would be the daughter of a neighboring nobleman who was well-connected and powerful.

Philip was decided to live a sad and solitary life, married to a woman that cared more about social activities and frivolities than caring for her husband. In order to avoid social ostracization or the retaliation of Dorothea's father, Philip continued to live in an unhappy marriage although deep inside he felt hollow and gloomy.

One day, as Philip was taking a regular lonesome ride at the boundary of his estate, he came along a Gypsy camp. Angered that the camp had set up on his property without permission, he approached them to demand their immediate departure. Philip was startled by a girl singing just outside the camp who appeared to be alone. She was a beautiful dark-eyed, raven-haired young woman who took his breath away.

Philip went to her and introduced himself, and she did the same.

Her name was Margo and she was the niece of a Gypsy woman that had taken her in after her parent's death. There was an immediate and powerful chemistry between them. He fell in love at first sight, and she felt the same. The pair saw each other secretly a few more times before eloping.

Defying common sense (and Margo's uncle as well), Philip secretly brought Margo back to live in the deserted gatehouse near the stables of Diddington Manor. It would be his and Margo's love nest, and for some time, it was.

But soon, Dorothea found out about the improprieties of her husband, and enacted revenge upon the hapless couple. Dorothea accused Margo of witchcraft due to her Gypsy heritage, saying that Margo had her husband under a spell in order to have an affair with a gentleman and steal him away.

The group decided that Philip would be a weakling who was too

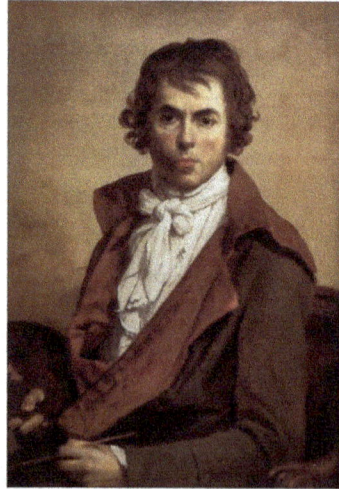

REPRESENTANTION OF
"PHILIP"

scared to confront Dorothea, and defend Margo from her lies. Dorothea's father would have had the power to take away his possessions and destroy Philip's reputation, so he stood silently at Margo's trial where she was sentenced to be burned at the stake (a deliberate, yet more tragic, historical inconsistency as English witches were usually hanged, not burned).

The lore then followed that Margo was executed horrifically, leading to Philip's eventual suicide. He would pass entire nights alone in a fit of despair, pacing the battlements of Diddington, until one morning when his battered body was found at the bottom of the battlements. He had thrown himself from the parapet in a final fit of agony and remorse. This was the story of Philip, a tragic mixture of fact and fiction created with deliberate errors in order to emphasize his fictitious nature.

The experiments conducted by the Owen Group were assigned to take place in a specific room, always with the same eight participants. It was felt to be safer for the manifestation to occur under these circumstances. The group decorated the room with furniture and keepsakes of the 1600s, arranging books, documents, maps, pictures and fencing foils (as they created Philip with a love for fencing). They even included old, framed pictures of the real Diddington Manor like the characters might have in their own home. For all intents and purposes, the room was decorated to appear as that of an Englishman named Philip, and created to appeal to a "real" individual that had died a long time ago.

The Owen Group met once a week to work on their ghost. They would discuss Philip and talk about his life and interests as if he was a real person that had once lived in that very room. They visualized him and even created a drawing of what he might have looked like. During every meeting, they would sit in a circle and meditate upon Philip in order to cause his apparition. But nothing happened.

The British parapsychologists, Kenneth Batcheldor and Colin Brookes-Smith, had been studying their séances and suggested a change of approach: that the group stopped being so somber and create, instead, an "atmosphere of jollity and relaxation." So, the members began to relax. Their focus turned from conjuring Philip to celebrating his existence. They started singing silly songs, eating candy, telling jokes, and addressing the table directly as Philip.

In the Fall of 1973, Owen conducted a session in which the table began to vibrate and allegedly moved of its own accord. "Philip" had finally arrived, and began to communicate with raps and knocks on the table. The raps and knocks were created by collisions of electrons, produced by a gravitic force of foreign origin. They had no physical mass, but were only gravitic waves colliding with the electrons of an area with mass, like a door, table, or walls.

Owen was ecstatic; the experiment had worked.

"Philip" answered questions consistent with the fictitious history the group had created, but could not reveal any new details. The Owen Group theorized that this manifestation came from their own collective unconscious.

Sessions with Philip continued for several years. The levitation and movement of the table was recorded on film in 1974, but efforts to capture Philip's voice on tape through Electronic Voice Phenomena, or EVPs, were inconclusive (although members of the group believed they had heard whispers in response to questions).

There are still many unanswered questions. What was Philip? Was it a collective unconscious projection of the Owen Group? What made it start, and what ended it? How was Philip manifested through nothingness into a sentient entity that communicated with the group for years? Why did it vanish one day for good?

The conclusion that Philip was a manifestation of the Owen Group is seen as both accurate and flawed simultaneously. They gave form and identity to the Philip entity, but how is it possible they could produce its essence? The conjuring of Philip was far beyond any of the participant's capabilities: the physical and neurological energy it would take to produce a thought-form simulation of a "ghost" could not be sustained by eight hundred people, least of all by eight. It should have drained nearly all the life force of the participants in order to create the bioenergy of this entity. But it didn't, and many other groups then went on to recreate the experiment successfully.

One group created and manifested "Lilith," a French-Canadian spy executed in France during World War II. The results were similar, but not quite as impressive as the Philip experiment. Yet another group decided to create a character from the future named Axel with even less impressive results than Lilith. Finally, one other group decided to create silly and improbable characters such as "Santa Claus" and "Silk the Dolphin," that were basically a waste of time. Philip became TSPR's greatest success.

If the complete results had been accurately reported and their findings reviewed by their peers/other research institutions, perhaps by now we would have a much more distinctive view of paranormal activity. The secrecy of such research done in first world countries such as the United States of America and Russia (especially during the Soviet Union) has considerably stalled our understanding of such phenomena. So, the importance of the Philip effect is paramount on the subject.

A few things should be taken into consideration regarding Philip, Lilith and Axel:

- These "artificial ghosts" only manifested within the presence of the entire study group

- They were created with benign, if not harmless personalities

- They were "contained" within the boundaries of carefully selected rooms for each of the experiments

- They were created and placed outside the sitter's own timelines

Therefore, the sitters created at least four lines of defense between themselves and their creations:

1) The "ghosts" were self-evidently imaginary, and that was known and believed by every participant in the group

2) Their fictional lives were over or had not yet begun

3) They were not characters created with the intention to harm the participants

4) Their ability to manifest depended upon the presence of the entire group, without which they could not be summoned

73

The following information was suppressed from Dr. Owen's notes and ended up, somehow, within the metadata of a certain intelligence agency working within the military that shall remain unnamed. It states:

A. The entity, Philip, was not a mental construct, but indeed an ethereal sentient cognitive bioenergetic transdimensional life form, or a spectre: **S**entient, **P**aranormal, **E**ntity, **C**ognitive, **T**ransdimensional, **R**adiant, **E**nergy. The S.P.E.C.T.R.E. is made of radiant bioenergy that travels as electromagnetic waves from adimensional and atemporal non-space to our universe.

B. The spectre used the combined thought-focus of the eight "sitters" as a beacon to triangulate the dimensional coordinates and ground itself in our universe. So long as the local elements of quantum mechanics and general relativity allowed its subatomic particles to resonate in our dimension, the entity could remain. When such "resonance" became unattainable, the entity had to return from where it came.

C. Entities are always temporary and seasonal. Temporary because they can not remain in our universe indefinitely and seasonal because their arrival in our universe depends on several basic conditions such as:

    *The stability of quantum weather (hyper relativistic quantum states of a given sector of the universe)

    *The right conditions of solar radiation (intense solar activity would impair the availability of transdimensional quantum vortexes to be successfully established)

D. Entities are attracted to emotional states, as moths are to the light. Because of their electron-based foundation, they are polarized into positive-prone and negative-prone induction, meaning: positive feelings (such as joy, kindness, caring and elation) are counterpoints to negative feelings (such as fear, rage and despair). Both polarized emotions connect the spectre to the "sitters." The spectres feed on these emotionally broadcasted biocharges.

It is also important to comment that the notes of this intelligence agency state:

1) The Philip entity indeed did manifest itself without the need of all eight members at certain times. Dr. Owen kept this, and other unprecedented occurrences, secret from the entire group.

2) Philip manifested in other sites outside of the designated room, and it came without being "called." This happened at least 5 distinct times and with only a handful of people from the experiment group.

3) Philip was not always benign. During the conjuring sessions with all eight members, it would behave as usual, but when outside of the designated room and with a significantly smaller group, it became abusive and mean (such as telekinetically hurling a book at a person).

4) Philip became linked to three individuals of the Owen group who all experienced emotional impacts of the entity within their private lives. It forced its presence at least once on each of these individuals when they were alone, but it was kept secret from the rest of the group to avoid any influences on the outcome of the experiments.

5) Philip was not just a fictional creation: something had embodied its identity and manifested itself in the place of this artificial person.

What the Philip experiment had done was not simply create a fictitious ghost. It had instead conjured a spectre into our dimension and gave it an identity and pseudo history to embody. Although the participants knew they were creating a fictional character, there was an element of fear nested in their subconscious minds that the entity was able to manipulate: a fear inherited by all humans; deeply suppressed genetic memories embedded within us from millions of years of primal cave dwelling ancestors who regarded bumps in the night as true threats of death. There is subconscious fear even when logic dictates there is nothing to be afraid of, and spectres feed on these extreme emotions. They hunt their prey based on both opportunity and availability, like all predatory creatures. Their most likely victims are:

A) Simple and average people regardless of their social status or intellect (though the less intellectual one is, the easier prey one would become).

B) Those in a state of transition: be it questioning oneself or an unsatisfactory relationship, anyone who is in the "in-between" of the old and the new.

C) People who are going through big changes, such as births, deaths, puberty, marriage, or significant decisions.

D) People who are experiencing great emotional turmoil, which may or may not be related to a transition or large life change. Entities thrive on emotional turmoil.

The Ouija was also used in the experiments, which must be taken into consideration. Using a Ouija board requires dedication of the sitters as they take their time in intimate physical contact. Such intimacy is not only perceived by the conscious self, but it has its own interpretations within one's subconscious self, as well. Regardless of gender or age, with glances and pheromones being exchanged across the planchette, repressed feelings can surface. These feelings can create emotional conflicts that the individuals would rather keep secret, and generate unexpected emotional turmoils that shouldn't exist in the first place. If this happens with adults, imagine the effect of this on adolescents.

An entity cannot do anything if one does not allow it to manifest. Keep in mind:

- It can not come and stay if it is not invited.
- It can be banished if the willpower is strong enough.
- It can not "possess" you if you do not allow it to do so. Your own fear weakens you and empowers the entity. Be brave and you will prevail. Have doubts, and it will be your downfall.
- It always lies. It has a sole agenda: feed from the bioenergies of our emotional output.

Governments know about this, but just like subjects of extraterrestrial lifeforms, cryptids, and astroarchaeology, they would rather keep it all under wraps for exclusive private domain and secret research, and let the population see it as far-fetched nonsense. Why? Because knowledge is power, and they certainly do not want the people empowered. Ignorance can be bliss.

In the next chapter, I will introduce you to other kinds of entities. The artificial thought-form and the artificial elemental.

Unknown by many uninformed people, the American Academy of Science recognizes parapsychology as a science, and the Philip experiment was a scientific project carried out by a group of researchers. The Philip experiment followed the scientific method, from making an observation to communicating results. The experiment tested different variables through Philip, Lilith, and Axel. The results of an investigation can only be considered valid if they are achieved during repeated trials of the same procedure, hence why these researchers spent several years working with Philip, Lilith, and Axel.

They used the K-W-L-H chart which follows the principles:

What I Know → What I Want to Know → What I Learned → How Can I Learn More

This Philip Experiment was done by scientific researchers in a controlled environment, but what if the same exact procedure was done by a group of people who called themselves a coven, during a so-called ritual? Say they received similar results as those that the TSPR achieved with Philip. What then?

Spiritualists, sorcerers, and even parapsychologists have their own points of view on what has been done, who has done what, and even how it was done, but they all share the same goal. All seek knowledge and some level of control over the paranormal.

There are no "occult powers," but simply unknown sources of power. Be one a mystic or a researcher, it is all about following an agenda.

From selfish reasons (such as earthly benefits of wealth, influence, [false] love, and revenge) to selfless goals (such as knowledge, understanding, [true] love, and caring), these are motivators that move the seeker into the realm of possible comprehension. These drivers open up the seeker's mind to the many theories and possibilities of the paranormal, providing a framework that may in fact lead the seeker to their sought knowledge.

This brings us to the dual concept of thought-forms and artificial elemental posits.

THE THOUGHT FORM AND THE ARTIFICIAL ELEMENTAL

A thought-form can be wrongly defined as a non-physical entity created by thought. In truth, the human mind cannot produce an entity, but it can feed the entity with information, allowing it to become anchored to our dimension. Human thought provides essence to it and gives it substance, though it is substantially ethereal. Thought forms are not created; more realistically, they are sustained by those giving it the means of its substance. Thoughts are powerful enough to provide a stabilizing anchor, based on the emotional output of the entity's targets. The thought-form establishes its identity from the very minds of its targets.

The thought-form in its true essence is an amorphous ethereal kind of S.P.E.C.T.R.E. that acquires an identity based on the information provided by its summoners (and future prey).

An artificial elemental is similar to the thought-form, with the distinction of instilling a principle or quality upon its target. While the thought-form would be seen as a predator, an artificial elemental would be more like a symbiont: an ethereal entity living together in close association with its target.

Dion Fortune, the British occultist, described her experience with a thought- form in her book *Psychic Self–Defense* (York Beach, ME, Samuel Weiser INC, 1999 – from her original book published in 1930):

Mrs. Fortune was upset by the betrayal of a former friend and was in a sleepy state lying in bed. She was thinking about revenge against that person when [sic] "I felt a curious drawing out sensation from my solar plexus, and there materialized beside me on the bed a large wolf. It was a well-materialized ectoplasmic form... grey but somehow colorless, and... had weight. I could distinctly feel it's back pressing against me as it lay beside me on the bed as a large dog might."

Let's examine how the thought-form could have accomplished that.

The brain is a complex organ that contains 90% of the body's neurons, and controls different portions of the body. There are three main structures at the base of the brain.

1) The cerebellum controls coordination, posture, and balance.
2) The medulla oblongata governs involuntary body functions like breathing and digestion.
3) The pons helps control the rate of breathing and relays signals between the cerebellum and the cerebrum.

The cerebrum is the large structure at the top of the brain. It controls motor coordination and interprets sensory information from inside and outside the body. The cerebrum is divided in two halves called hemispheres. Each hemisphere contains four lobes that perform specific functions:

- Frontal Lobe – Regulates voluntary movements and is involved with decision making and problem solving.
- Temporal Lobe – Regulates memory, emotions, hearing and language.
- Parietal Lobe – Processes sensory signals from the body.
- Occipital Lobe – Involved with sight and visual memory.

The spectre manifested beside Dion Fortune by following the signals broadcasted by her intent of revenge upon her traitorous friends. Once it located the signal and established a presence, a "rapport" process was initiated.

The dendrites in Fortune's body received a broadcast by the spectre that was then turned into an electrochemical message. It told Fortune's occipital lobe that she was actually seeing a wolf-form at her side.

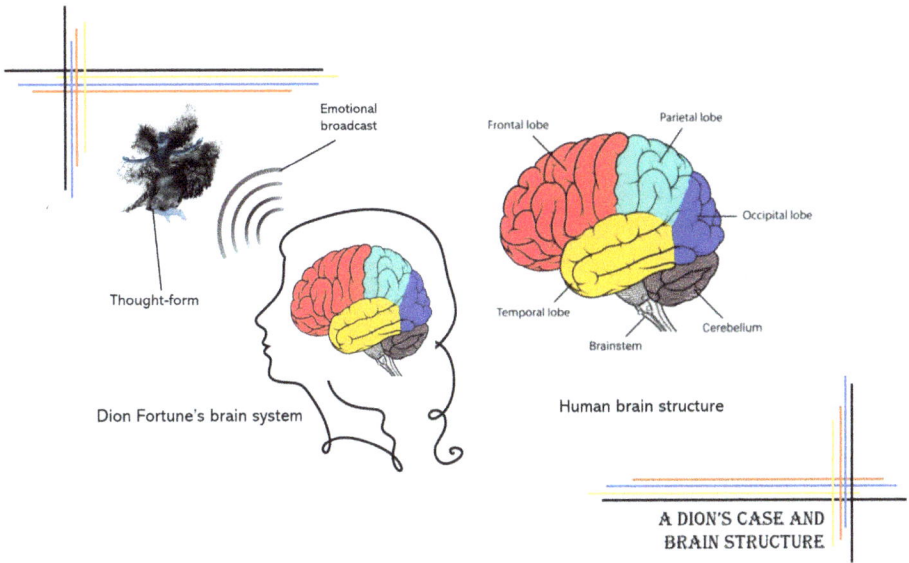

Emotional broadcast

Thought-form

Dion Fortune's brain system

Frontal lobe

Parietal lobe

Occipital lobe

Temporal lobe

Cerebellum

Brainstem

Human brain structure

A DION'S CASE AND BRAIN STRUCTURE

Then, another message was sent to Fortune's frontal lobe to accept the wolf-form as real, though not an authentic wolf but an "ectoplasmic" form of one. Her temporal lobe received the message of fear and danger — and Dion believed that was her negative thought taking material form. The parietal lobe received a signal registering the physical weight of a wolf (although no wolf was indeed there). The temporal lobe associated the apparition as "evil" and instructed to not obey its commands. Its nature was then established based on her knowledge of the subject, and her conclusion was that of a "lupine apparition," or an ectoplasmic werewolf. So, she defused the ectoplasmic entity, which truly wasn't ectoplasmic at all, but solely a sensorial implant placed by the spectre.

Fortune fought with her willpower and won, banishing the

creature away, and actually severing her link with the spectre. Dion Fortune truly believed that she could produce a thought-form under the following conditions:

1) The enabling actor needs to be in the condition between sleeping and waking.
2) Their mood has to be one of brooding and highly charged with emotion.
3) They have to make an invocation of the appropriate natural force.

Fortune never considered that she had been used by the spectre, but it used Dion's own brain against herself and gorged upon her emotions.

As far as artificial elementals go, Alexandra David-Néel introduces them in her book, *Magic and Mystery in Tibet* (New York, New York University Books, 1958, actualized version of her original book published in 1929).

Madame Alexandra David-Néel (1868-1969) was indeed durable, living until the age of 101 years. She was a Belgian-French scholar, traveler and published writer who worked with thought-forms and elemental entities, predators of bioenergies, from the ethereal elsewhere. She was also a Buddhist that spent over fourteen years in Tibet, studying with swamis, hermits, monks and lamas. There she learned of "magical phenomena" such as:

- Messages carried on the wind to faraway (telepathy)
- Enchanting a knife so the selected victim would use it to commit "suicide" (telepathic mind control)
- Tumo (the ability to generate high temperatures in the body, or thermogenesis)
- Tulpa (the ability to "create" or call-upon thought-forms and elementals)

We will be discussing the Tulpa phenomenon specifically with more detail. David-Néel stood many months in solitary meditation and made and participated in certain rituals of invocation until she could call upon her own Tulpa. It came in the form of a monk, exactly as she had envisioned it to be: short and fat, bald, and an innocent and jolly type.

To visualize her Tulpa, David-Néel had to go through a long and difficult process of concentration until she could actually "see" it. The "mind monk" was how she thought about her own Tulpa. No one but her would be able to "see" it, since she saw it with her "mind's eyes." She traveled with a party of aides, and it was a sight to see her talking apparently to herself.

In the beginning, David-Néel's Tulpa remained as she imagined it, and it obeyed all her commands. But soon she realized that behind its benevolent stance, there was something a bit more sinister within it.

Small events started to occur during her journey. People started to get hurt: nothing serious, but nothing to be taken lightly either.

David-Néel discovered that as she summoned the monk, it would take its own time to manifest and became independent. It's appearance started to change. She meditated on its features as bald, short, fat, chubby-cheeked, clean-faced and jolly, but it started to materialize as leaner, wearing a dirty tunic, with facial and body hair beginning to show. As it grew even leaner, with longer hair and beard, it's face assumed a vaguely mocking, sly and somehow malignant look. It would manifest itself without being summoned. It's pranks became harsher, more dangerous, more cruel. It turned bold and downright wicked, escaping totally from David-Néel's control, if it ever had been under her control at all.

The Tulpa wasn't the thought-form she thought it was, but indeed an elemental force upon which she imposed artificial elements of identity.

Yet, David-Néel insisted on her thought-form explanation, declaring that she saw the Tulpa's change in behavior to be inevitable.

According to her, once the Tulpa is endowed with enough vitality to play the part of a real being, it tends to free itself from its maker's control. It is clear that she still believed that the Tulpa was "created" by her. The Tulpa is not a fabrication, but an independent entity seeking someone to feed upon until it becomes strong enough to act at its own will. David-Néel continued, saying:

> "According to Tibetan occultists, this separation happens nearly mechanically, just as the child, when his body is completed and able to live apart, leaves its mother's womb." Yet another excuse supporting the idea that she had created the Tulpa and not just been used by it.

Continuing her odyssey with the Tulpa, David-Néel relates that its very presence became so disturbing that it turned malevolent in nature. Trying to "dissolve" it took more than six months, though this was an impossible task. The only thing she could have done was to banish it back to where it came from, but she continued to insist on the Tulpa being her creation. It was a task that required half a year of hard struggle with her mind; the creature was tenacious to keep its life.

The local Tulpa-lore states that not all Tulpa are indeed "created." They say that a Tulpa can also appear spontaneously. For instance, if a traveler was passing through some sinister tract of country, it would come along to "haunt" them, or even try to possess them.

As an example of a Tulpa's spontaneous apparition, we have the case of Reinhold Messner, who made a solo ascent of Mt. Everest in 1980. Messner described an invisible companion who climbed beside him, as noted in the book by Jon Krakauer, *Into Thin Air* (New York: Villard, 1998). Alpinists see and hear many strange things, but also suffer greatly from fatigue and lack of oxygen. But when physical evidence becomes available, the "imaginary" becomes quite palpable. Not all Tulpas are malignant in essence but, nevertheless, they will sap one's life force at any given time.

Could the Tulpa be created unconsciously? Well, I would not say "created," but "summoned" perhaps. When thinking that they have "created" the Tulpa, its summoners do not realize they have no power over it at all. The Tulpa obeying commands is a ruse performed to gain the confidence of its summoners and create strong links between the entity and the human(s) linked to it. The summoner of a Tulpa does not aim for a fixed result, since the outcome is as unpredictable as the phenomenon itself. In some ways, the Tulpa is like a parasite that feeds from the emotional output of its summoner. In order to gain control of the summoner, the Tulpa will be a deceiver all the way.

Tulpas are elemental entities, amorphous ethereal sentient cognitive life forms that can be from our own universe or from other realms. But Tulpas are also artificial elementals as well. How so? Because its summoners will give it an identity that it can latch onto and claim, falsely, as its own. Hence an elemental entity becomes an artificial elemental entity.

David-Néel describes instances of sorcerers being killed by their own thought-forms or artificial elemental creations. Remember, the ethereal parasite is feeding from one's life force. The point to be made here is that, the more you think about it, the more dangerous it will become. By empowering it, it establishes a mind-link over its hapless victim(s).

The warning here is to not dwell upon the superstitious side of the paranormal because it is primal and leaves you unprotected. The "summoning," "invocation" or "connection" (or whatever name you decide to call the action of connecting with an ethereal lifeform, be it a miasma, a spectre, a thought-form, an artificial elemental [Tulpa], or any other non-corporeal parasitic predator), will always open a Pandora's Box that will be difficult to close once opened.

While many would expect these ideas to come from Christian tradition, it is a worldwide belief that the devil/evil/wicked forces cannot go where they are not welcome.

Dion Fortune and Madam Alexandra David-Néel found that even when playing by the rules (according to traditions and lore), it is an unfair game. The game is rigged, my friends. Its rules will always change in favor of the house, and we are always playing in the house of the occult.

CHAPTER 09

THE FAITH AND THE DEMONIC SIDE OF THE OUIJA

Among the Native Americans, the Ouija board had its equivalent: the Katsinam. The sacred Datura, a hallucinogenic plant, tells its takers of the Wheel of Divination, and the Bridgeway to the Afterworld (land of the spirits). The Katsinam was a doorway into this world.

Katsinas are usually dolls similar to Spirit Dolls. However, the Hopi Manitou-Katsinam is different. The Hopi Manitou-Katsinam is said to be linked to the Hopi stone tablets of Techqua Ikachi, the Four Arms of Destiny. It was used by the Hopi Snake Clan as their sacred Spirit Wheel of Divination.

When the Manitou-Katsinam was put under the dominance of Orion in the winter skies, flanked by Taurus (the Bull) in one site and Canis (the Great Dog) in the other, the stars would appear in the entryway of the kiva roof (a sacred underground room) to allow the spirit of the Manitou to divinate in the matters of the afterworld. The spirits would come and talk through the doorway to the underworld opened by the Hopi Manitou-Katsinam.

One must beware of the tricksters, the Skinwalkers that feed from people's souls, or so the lore of the Native American warns. In the end, we see nothing but miasmas, spectres, thought forms, and/or artificial elementals as the culprits of harm.

The Skinwalker, ever-elusive, intrusive and insidious, was the Native Americans' foe. The Skinwalker would deceive, lie, and cheat as long as it could influence its prey and feed from their fear, rage, and despair. It could be beat, but a great deal of willpower, courage, and determination are necessary to win against such an evasive adversary.

Such spirits were once called upon through the Wheel of Divination to answer important, if not critical, questions. Anthropologists find veracity in the myth of the Skinwalkers. Hidden over thousands of years of Native American lore, the Skinwalkers are not simply the product of manifestations of the denizens of the Natives' collective consciousness.

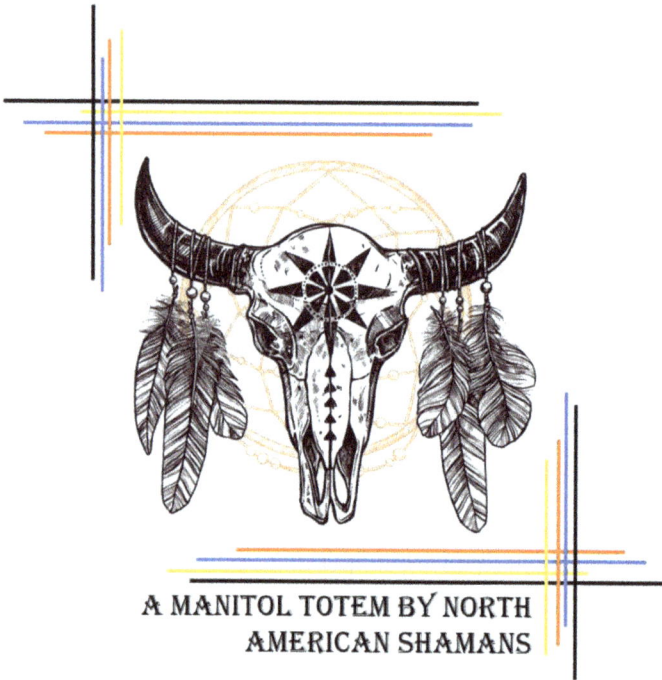

A MANITOL TOTEM BY NORTH
AMERICAN SHAMANS

The kivas housed the Manitou-Katsinam and through it, one could reach the Spirit World. Can you see the correlations between these elements and the Ouija board?

Half a world away, in the middle East, we have the legendary Jinn (or Genii).

The Genii are spirits of elemental forces which allegedly grant wishes to the unsuspecting victim, giving them the illusion of control, when in fact they were the ones being manipulated by the entity. The oil lamp was deemed the "magic lamp" because one could conjure the spirits in its flame, as the old tales tell.

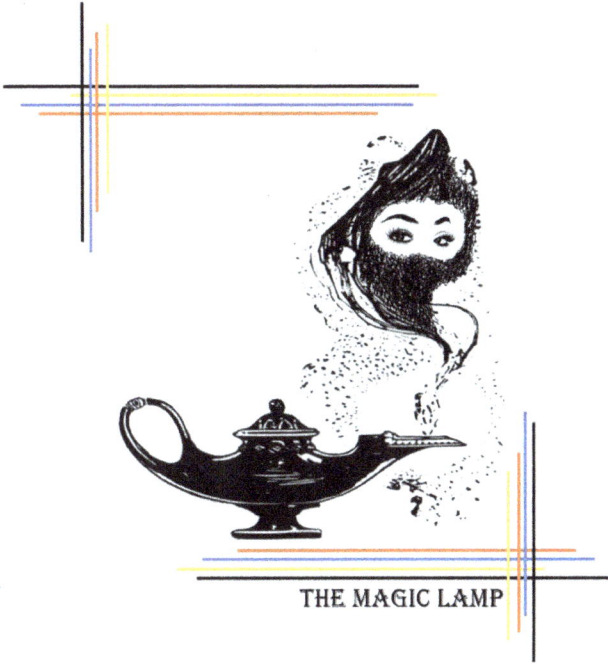

THE MAGIC LAMP

Europe had the Fey, the legendary beings of folklore and romance: usually diminutive humanoid creatures that had alleged magical powers. Even in modern day, popular culture refers to them as "fairies," the people of Fata (the goddess of fate). What few know is that the original Fey were not considered beings, but spirits of the underworld. The "token" to call a Fey spirit was a piece of gold, or any golden item. The Fey were to tell divinations to their summoners.

The Fey, like the Trickster and the Genii, were ethereal entities. All of these entities were nothing but the arcane lore manner of designating spirits of the so-called Underworld, the realm of the dead. The Manitou-Katsam, the oil lamp and the golden objects perform the same role as the Ouija board does upon summoning spiritual entities.

The Trickster, the Fey, and the Genii are also known by their ability to "possess" people. By spiritualistic definition, possession is usually the control of a living being by an external ethereal agent.

REPRESENTATION
OF "THE FEY"

Such a possession might be by a spirit, a godling, or the soul of someone dead. There is a clear distinction between a spirit and the soul of someone dead; a spirit is a foreign entity. There are two types of such spirits: light spirits, if they have benign nature, or unclean spirits, if they have malevolent nature.

Possession allows the spirit to interact in the physical world. It remains an essential practice across the shamanism of North America, the Aborigines of Australia, the African-related ritualistic, and the faithful saints of European belief. Even contemporary Pentecostal and charismatic variations of Christian faith worship these spirits, actively seeking possession by the Holy Spirit.

Collin de Plancy's *Dictionary of Demonology* (New York Philosophical library, 1965) and Martin Ebon's *Exorcism: Fact, not Fiction* (New York: Signet Books, 1974) relate: "Evil spirits" is a flexible term that can be used for any "supernatural-related element." They could be the spirits of wicked people who had lived, or even hate-filled, malevolent entities that never lived (such as demons, devils, or fallen angels). Each kind of "evil spirit" would have its own "personal characteristics." For instance each individual "possessed" would exhibit peculiar mannerisms, special knowledge, likes or dislikes, and even allergic reactions.

REPRESENTATION OF POSSESSION

The demonically possessed are known as demoniacs, and they usually display a preternatural knowledge of past, present, and, allegedly, future events. They may speak and understand foreign and ancient languages not understood but by very few scholars and surely unknown by the victim. They are repelled by religious icons and rituals, sometimes showing markings upon their skin between the exorcism's rituals.

According to the dictum of Christianity, in what is known as the Miracle of the Swine, Jesus expelled a plethora of demons, called a legion, from the Gerasene demoniac. The demons were expelled into a herd of pigs, which went into a desperate frenzy and drowned themselves in the sea.

Note: Parapsychology explains some of the phenomena related to the so-called "Demonic Possession Syndrome" as a hyper-inducing of hormonal peptoids, steroids and adrenocorticotropic (and adrenocorticosteroids) produced from the adrenal cortex and other sites and moved via bloodstream to areas affecting the physiological activity and the metabolism of the victim. Some of the more common phenomena are:

A. Xenoglossia – Capacity to access information or languages of ancestral cultures worldwide by tapping into genetic memory found over the so-called "junk DNA." As controversial as it might be, a great number of researchers accept the possibility of genetic memories as a form of DNA-enabler carried through biochemically-induced encoded data over generations of ancestry.

B. Sansomnism – An increase in magnitude of physical strength and stamina by hormonal surges of adrenaline.

C. Telekinesis – Redirection of graviton fields induced from the pineal gland producing a levitational field over specific matter.

D. Thermogenesis – Capacity to accelerate or slow air molecules in a site to produce heat or cold, through the use of telekinesis.

E. Pyrogenesis – The use of telekinesis to increase the heat of molecules frictioning against each other, producing ignition of flammables.

F. Dermatogenesis – Alterations of skin tissue through the change and configuration of skin fibers at the molecular level. Facial features can be altered (impermanently) through muscle contractions.

G. Telepathy – Connection between the parietal lobe of the brain linked by transmission and reception of bioelectrical impulses in Hertz frequencies, produced by the neurons of individuals.

H. Neuromotor Chronic Induction (N.C.I) – Creation of temporary muscular alterations that could allow the victim to assume positions a human body would not normally be able to perform (e.g. extreme curving of the spine for instance).

I. Meta-animism – Through the use of telekinesis, the ability to give apparent "vitality" to lifeless objects (e.g. a chair that moves as though it is alive).

J. Neuro-induced Hallucinations – On which the "possessed" will be able to produce realistic illusions to torment his/her prey.

Medieval and Renaissance authorities believed that demonic possession resulted from witchcraft or an "evil breed" when, perhaps, they were haplessly suffering from some psychotic illness or even being victimized by ethereal predators.

The very controversial, but widely read, author, proclaimed exorcist and former Jesuit, priest Malachi Martin (1921-1999) believed that evil spirits concentrated on perverting the will, not the body, and that they must have aid of the very victim in order to succeed. In his book, *Hostage to the Devil*, Malachi Martin wrote, "At every step and during every moment of possession, the consent of the victim is necessary, or possession cannot be successful" (New York: Perennial Library, 1987).

The caveat is that the consent of the possessed can be subtle, almost unconscious. Or alternatively, it can be as direct as an agreement between the afflicted and the entity.

However, the Roman Catholic Church does recognize the rare occasions in which, according to Donald Attwater's *A Catholic Dictionary*, "God seems to allow even the innocent to be exposed to the physical violence of the Devil" (New York: Macmillan Company, 1958 – Page 390).

Priest Malachi Martin describes the event of a possession as a four-step process:

1. Entry Point - The point at which evil spirits enter an individual and a decision, however tenuous, is made by the victim to

allow that entry or not, whether by consent, fear, or deceit.

2. Erroneous Judgements "by the possessed in vital matters, as a direct result of the allowed presence of the possessing spirit and apparently in preparation for the next stage."

3. Voluntary Yielding of Control - The point at which the possessed person yields to a force or presence, thus losing control of their free will, decisions and actions.

4. Perfect Possession - At this point, the force or presence has complete control over the human victim. This is not the swearing, floating, and paranormal phenomena kind of possession, but a functioning, apparently normal condition that leaves the person devoid of humanity. Such perfect possession would explain how some people, when possessed, act absolutely contrary to their normal selves and commit savage and barbaric acts against others.

From *The Week* magazine, March 9, 2018, Volume 18, issue #863, Page 6:

"Good week for…employment opportunities, after the Vatican launched a major training program for exorcists. Church officials say the need for exorcisms has tripled over the past years, due to the use of fortune-tellers and tarot readers who open the door to the devil. Inside sources also denounced the danger of selling Ouija boards as toys for children, teenagers, and adults, that could become prey for demons. Or so they say."

On the same token, reality can be more amusing than fiction. A Pennsylvania church known as World Peace and Unification Sanctuary held a "gun blessing" commitment service for its congregants, in which congregants were fitted with bullet crowns and AR-15s. The Sanctuary believes that the Book of Revelations' prophecy that Christ will return to Earth and rule with "a rod of iron," clearly refers to the popular assault rifle. "The rod of iron is the AR-15, in today's terms," said church leader, Tim Elder. Nevertheless, rumors said that Elder used the Ouija board and was enlightened by "spirits of light," which gave him such ideas. What to make of this?

In their book *Wisconsin Lore*, authors Robert E. Gard and L. G. Sorden claim that even among those who were born and raised in Wisconsin, where "there are more ghosts per square mile than any state of the nation," not many people have seen or had a personal paranormal experience (Madison: Wisconsin House Ltda. 1962). Yet, it is mostly young people who dare meddle with the Ouija board in graveyards, haunted houses, and even satanic churches, with mixed or inconclusive results. However, when you seek evil, you have great chances of finding it more often than not.

Traditionally, Ouija sitters held distinct and often unconventional social, spiritual and political opinions. Seen as eccentric, they usually have lives that are not exactly monk-approved. In fact, these sitters may even be seen as downright sinful. It is simply human nature to be drawn to the sinful. Take, for instance, college campuses, especially in the United States, which are fraught with sex, alcohol, drugs, and more sex. Such places would be especially attractive to demonic activity.

They all should have been overrun by evil spirits by now, no? Somehow, they are not. Is the introduction to an occult element, such as the Ouija, the catalyst that would transform the situation into something more insidious?

We need to consider the distinct perspectives of demonic activity. Their introduction to the occult was the Ouija board.

Take the following as an example of a possible introduction to the occult: A group of relatively normal young people decide to gather and hold a series of séances. Motivated by curiosity of the trends launched in social media, books, movies, TV shows and urban legends, they seek the thrill of the supernatural.

So, contact! They contact spirits who allegedly never lived and receive a series of long and very dull messages. Their interest wanes. They want the beyond, the occult. The idea appeals to them. Some are frightened, others are aroused by it.

Once contact is made, the outcome of it is anyone's guess. It could be nothing at all...but it could be something.

These "spirits" will present themselves as benign, spirits of the dead, spirits that never lived before, elementals, angels or even demons. Through this connection, they are consuming the life force or emotional output from the sitter, upon which they thrive.

No matter the spirit in question, know one thing: get to know the true name of the entity, as this secret name is the key to the essence of its power. The name will not only identify the entity but also provide the entity's true intentions. To know it by its true name is, to some degree, to control it. Few know that by some unknown cosmic ruling, as the lore says, an entity is compelled to reveal its true name once they are commanded, and not merely asked for, by its summoner, or in this case, the sitter.

Using the Ouija board, or holding a séance, opens a sort of psychic beacon to entities over atemporal/adimensional non-space, that are eager to feed from the emotional output, if not the very life force, of their summoners.

How do we categorize such entities as good, evil, or harmless?

We cannot because at the core of this question, there is only

REPRESENTATION OF EVIL SPIRIT

one point to consider: be them benevolent or wicked, these entities are ethereal predators, and we are their prey.

These beings can consume joy or goodness, just as they can savor fear, rage, and despair. Whatever emotion it might be, positive or negative, it will be consumed. We become nothing more than a resource to them.

By having faith and believing in them, you also empower them.

Angels and demons are not but distinct sides of the same coin. Astral mischief will happen, be it on a positive or negative note and even when, apparently, nothing happens, or nothing seems to be present, it doesn't mean nothing happened, or that something isn't, in fact, there.

There will be signs.

Seemingly irrelevant small power shifts in the electricity. A dog, cat or other domestic animal acting distinctly.

Pay attention to them.

CHAPTER 10

THE OCCULT, THE OUIJA, AND STRANGE SIGNS

The true midnight isn't at 00:00. Not for these matters.

The true witching hour isn't midnight, but later.

The true midnight is 3:00 AM, the time between 00:00 and 06:00 AM hours, or halfway through the late nighttime. 3:00 AM is the Soul's Midnight. That is their time.

It would require serious statistical work to verify all the claims about 3:00 AM but the idea of such happenings occurring at this specific time is certainly widespread among the researchers of the phenomena.

It is called Soul's Midnight because of the many deaths and suicides that occur within the sixty minutes between 3:00 AM and 4:00 AM almost everywhere. There are exceptions of course, but usually, when such events occur, they would occur within the hour of the Soul's Midnight, or so the lore says.

Old soldiers who faced combat in long-past world wars claimed that the worst watch to be on was between 3:00 to 4:00 AM because they believed that it was when one's soul is at its lowest.

Nurses whisper that 3:00 to 4:00 AM is the "Death Hour" because it is the most common time for patients to die. It could be for many reasons but among nurses, it is believed that as humans are often in their deepest sleep at this time, things can turn very delicate. Hospitals are places of birth and healing, but they are also places of death and

suffering. Others call this hour the "coldest part of the night," not in reference to temperature, but to things that move coldy within the umbrae of the shadows.

Jim Krane in *Charles "Chuck" Conte, a Tenacious Detective*, wrote, "I can also state from personal experience of signing search warrants, that the police still like to raid drug dealers at 3-4 AM, as they figure they will be at their low ebb then and less likely to put up resistance" (Star Ledger, Newark, August 31, 2000).

Thus, be aware of the sign when the clock reads 3:00 AM.

Animals and Wave Reception

A wave is a repeating disturbance that transfers energy as it travels through matter or space. The world around us is full of many different types of waves. Sound waves, thermal waves, light waves, and radio waves surround us most of the time.

An important property of waves is that they transfer energy, but not matter from place to place. Of course, we can feel some of these waves ourselves, but we never feel them as distinctly as animals do. Animals always know when the weather will change or when a natural disaster is on its way. Animals can sense wave patterns in a way that we – humans – certainly can not.

For a frequency that is in hertz (hz) and a wavelength that is in meters (m), the speed calculated will be in meters per second.

In the formula below, V represents speed, F represents frequency, and lambda (a) represents wavelength.

$$V = F \times a$$

Consider a sound wave with a frequency of 160hz and a wavelength of 2.13. The speed of the sound wave is 340, 8 m/s.

A dog will hear this long before we do. Dogs and cats can see far

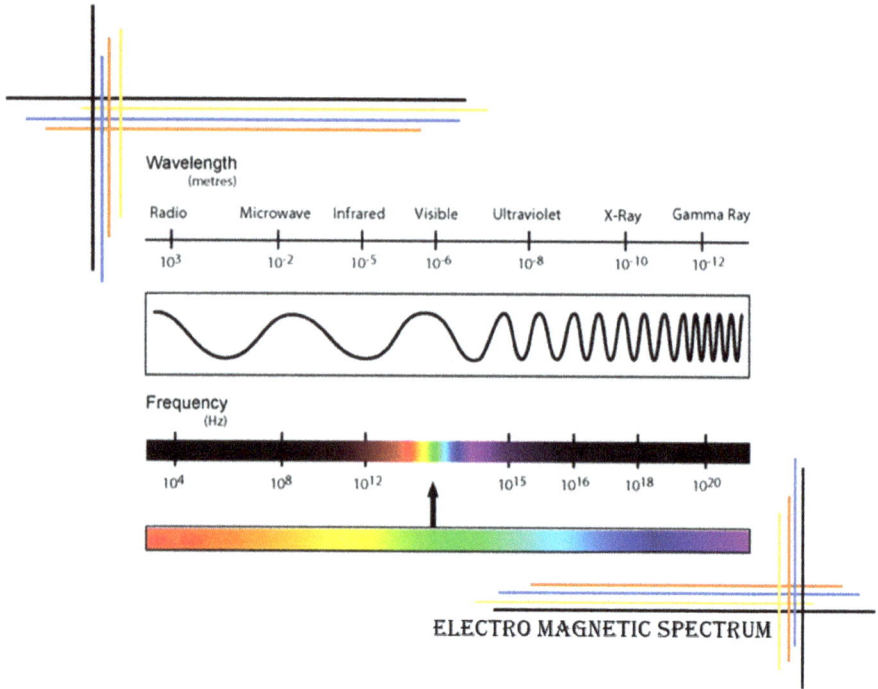

Wavelength (metres)

Radio Microwave Infrared Visible Ultraviolet X-Ray Gamma Ray

10^3 10^{-2} 10^{-5} 10^{-6} 10^{-8} 10^{-10} 10^{-12}

Frequency (Hz)

10^4 10^8 10^{12} 10^{15} 10^{16} 10^{18} 10^{20}

ELECTRO MAGNETIC SPECTRUM

better than we do as well. They can see what we can not.

Pay attention to the signs of strange behavior of dogs, cats, and other pets. They know what is really there.

Quantum Teleportation and the Occult

Certain nations use quantum teleportation for secure communications. Quantum teleportation is when given particles are metamorphosed into a quantum state and transmitted to join particles elsewhere using different kinds of entangled particles as the carriers. There are nations on earth performing Quantum Experiments at Space Scale (QUESS) with entangled photons and with practical conclusive results.

Quantum teleportation is viable on inorganic molecular compounds as well; small things such as pens, keys, a hairbrush or a remote control could, in theory, be demolecularized and quantum teleported to another area. The amount of necessary energy to produce such an event will depend on the source of the displaced power, either having it on storage or accessing it from a limitless nexus. Even though it seems unlikely, it is indeed viable and possible.

The point is not to demolecularize objects but to transport them through a Quantum Spatial Displacement Vortex Portal (QSDVP).

Parapsychology calls it the apportation effect.

Objects can be easily misplaced, and this is the most common explanation for their disappearance from the site that one was "almost certain" they left there. Sometimes the objects turn up in the most unexpected places or even in a place we had looked for it before to find nothing there. Even in the case of Alzheimer's disease, a degenerative brain disease that is characterized by progressive mental deterioration and memory loss, it is not the explanation for these "misplaced" objects. People of all age groups suffer from such experiences and just ignore them. What else would they do? Most people do not even know of QSDVP's.

When items disappear once in a blue moon it is nothing but a brief inconvenience. But when such occurrences happen more and more often, something is certainly going on. It could be pranks, or it could be something far more sinister and dangerous. It may seem harmless for items to temporarily disappear only to be found again, later on at some other place, or even at the same place they originally were. But what about strange things that should not have been there at all? Do not rush to conclusions, but do not ignore the unprovable.

Another sign to be aware of: how often things are missing and how long until they are returned, if ever.

Thermal Changes and Microclimates

Microclimates can be formed within a room of a home. Even with air conditioning, rooms are not equally distributed thermally.

Slight differences in temperatures may happen within regions of the room (usually at the ceiling or down to the floor) with different pressure points and humidity levels. Thermal microclimatic fronts often form in the least expected places.

The arrival of an external agent such as a miasma or a specter can move masses and cause a distinct shift in temperature.

A vortex portal would make atoms and molecules interact in a distinctive way. Such an opening would rapidly move particles and suck them into the vortex absorbing all heat energy, leaving a void of slow-moving particles in its wake – so-called "cold spots."

When searching for paranormal activity, one must seek "cold spots" because they will mark the entry point of a vortex portal. Atoms at the site of a vortex move very slowly. The slower they move, the colder the site will be.

Other times, unaccountable heat sources may also be detected. An entity could be absorbing thermal energy by kinetic displacement. Faster-moving particles have more kinetic energy than slower-moving particles. When particles collide, energy transfers from the faster particles to the slower particles through thermodynamic transfer.

We face such thermal changes at our homes daily but never give too much attention to it.

Beware of extreme or unexpected thermal exchanges. You might not be aware that external factors are influencing your environment. Spectres are not figures of imagination. They are a clear and present danger to be aware of.

The Occult and Electrical and Magnetic Energy

Some objects can affect other objects from a distance due to the force field that exists around them. A force field is a push or pull in the given region around an object. Electrical and magnetic energy are both results of such fields. These forms of energy are related to one another through ethereal electromagnetic charges within their quasi-physiologies.

The innate bioelectromagnetic fields surrounding such entities are usually a tell-tale signal of how they are above or below our visual optical spectrum, rendering them virtually invisible.

But even they follow the properties of electromagnetic wave characteristics. They can carry energy through matter or through space by using a medium, the available property allowing such movement. Thus, they can travel to almost all states of matter.

Such entities can broadcast within the whole electromagnetic spectrum, discharging X-rays, microwaves, infrared rays and others. An entities' ability to manipulate radio waves is what allows them to produce Electronic Voice Phenomena (EVPs). The "voices" one hears in paranormal recordings are synthesized into audio signals in frequencies and are not audibly perceived by average human auditory means.

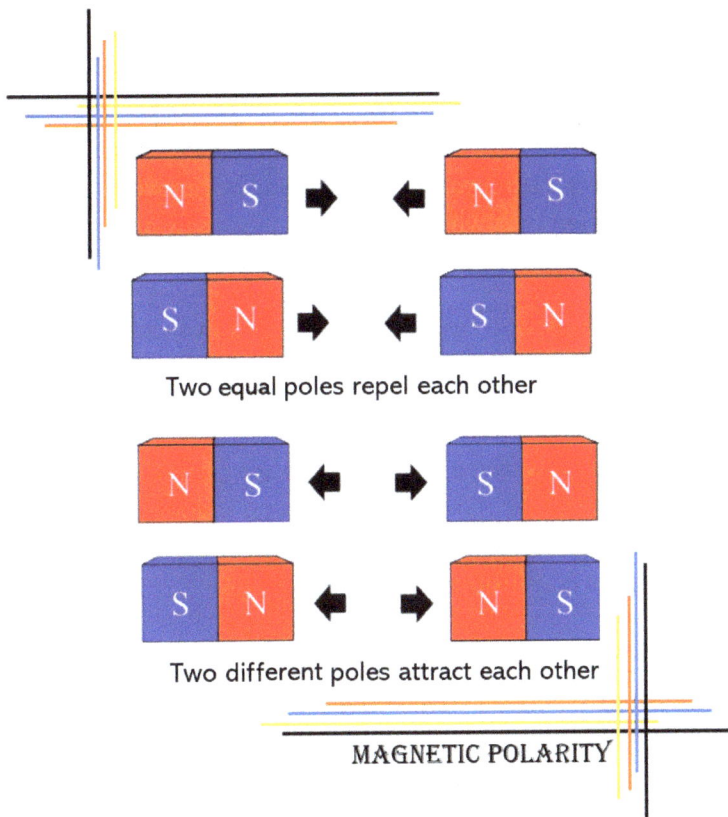

Two equal poles repel each other

Two different poles attract each other

MAGNETIC POLARITY

Types of Energy to Expect

- Electromagnetic
- Electrical
- Magnetic
- Thermal
- Radiant
- Radiant and Thermal (producing)
- Kinetic (due to an expansive wave blast)

Take note of the sign of unusual and unlikely electromagnetic discharges. It would be wise to have access to a spectrometer to measure the whole spectra, and a spectroscope, an instrument that produces spectra, especially of visible electromagnetic radiance. Spectral imagery can usually be found under such conditions.

Visionary Illusions

How many times have we spooked ourselves believing something was there when indeed there was nothing? A sudden movement caught in our peripheral vision that when looked upon, we saw an empty room. Was it an illusion?

Illusions are misconceptions of vision, or a misleading visual image. There are several reasons for having such illusions.

Sometimes, there are also mirages, not illusions, such as those that appear in a pool of water, a mirror, in the sea or in the desert. These mirages appear in the most unlikely places, indoors or outdoors, and can be seen in the daylight or the nighttime. These are the rare, metaphysically-induced illusions.

The human eye works together with the brain to help us see. The eyes let in the light and send signals to the brain, and these signals create the images that we see.

In the regular process of vision, light goes into the eye, first passing through the cornea, which is the clear layer outside of the eyeball. From there it goes through all the visual optical processes, transmitting external input to be processed by the brain.

When it comes to the production of illusions, an entity broadcasts a series of wave particles encoded with the patterns it wants to convey to its prey into the electromagnetic spectrum.

These signals are sent directly to the occipital lobe of the brain, which is involved with sight and visual memory. The encoded signal (with the patterns chosen by the entity) will be processed almost simultaneously by the two parts of the occipital lobe. Sight will provide the prey with the image the entity chooses to impose to its victim, and a visual memory will establish cross-referencing in order to recognize, or, at least interpret, the image reflected, even if there is nothing there to actually be seen. The image will always appear very real to the victim, but it is an illusion only. No one else will be able to see it, unless the entity has also intercepted others' occipital lobes.

However, electronic equipment will be able to record the approximated configuration of the images (invisible to the naked eye) because they are made of electromagnetic wave particles, and therefore, detectable.

Still, no eye or piece of equipment will be able to detect the true biometrical form of the entity, which is not humanoid in shape. The apparatus can detect the echo of the signal sent from the entity to the occipital lobes of its victims because it is of an electromagnetic wavelength.

The entity is amorphous and shapeless. It does not have or need a humanoid body configuration, but it can transmit any type of body configuration, humanoid or otherwise, if it chooses to do so. Nevertheless, electronic equipment can capture its shapes in the form of amorphic images and spherical orbs.

So, sometimes our eyes can and will, inevitably, see things that are not actually there. Things that only one can see, and others cannot.

The sad truth is that if psychology and psychiatry were just a little more observant and innovative, they would realize that not all cases of apparent psychosis are indeed what they seem to be. This does not mean that all neurotics and psychotics fall under such a rule. According to mimetism in evolutionary biology, what seems to be a leaf will not always be a leaf, just because it looks like one.

This is a cautionary metaphor to avoid assumption when it comes to the paranormal. Simply put, to assume makes an ass out of you and me. The sign to be wary of: visions that only you can see.

The Ouija and Strange Signs

The monotony of daily living can be tiresomely boring, especially among young people who think they are invincible and will live forever. Young in this case can refer to the age or the mind. Even if one's youth does not show externally, naivety and youth of the mind will certainly show.

For a person young of the mind, the world becomes again full of wonders and new possibilities, driving one to see way beyond the sphere of boredom of one's life. One aims for action, adventure, and even the thrills of the unknown. The occult offers unlimited possibilities, and the Ouija board becomes another venue of dabbling with the matters of the unknown.

The Ouija acts as a focal point that leads to a catalysis, meaning a change in the energetic configuration of the site, as it changes from a simple piece of decorated wood into a literal catalytic converter connecting feedback. There are plenty of listeners to these broadcasts, but there are many broadcasts, and it all depends on who or what is broadcasting.

Not all humans have the very same neural pathway systems, but physiologically and anatomically, they will each look almost the same. The difference isn't at a material level but has to do with the individual essence of one's true-self, or as people simply call it, the soul.

These entities can potentially access any soul they want, but they are selective in their choices. They are not looking for strong ones, because they know they will not win. So, they prey on weak souls instead. But even those can be surprisingly resilient. Human souls are unpredictable, and only time and circumstances will tell the ending results of such endeavors.

Humans are only prey if they decide to be. The human potential is surprisingly awesome and even these entities recognize it, despite meddling with our species. Entities are usually confident in the fear syndrome to manipulate their victims into submission. When one faces the entity with one's true faith, the wicked things have no other choice but to leave in defeat. Sureness and confidence are unbeatable traits to use against such entities. Remember, any doubts will be the weak links of your chain of strength.

Know this – the Ouija is a door only if you make it to be one.

Spirits are nothing but ethereal life forms. They have no control over you or even the environment. By themselves, they cannot produce gravimetric fluctuations to produce telekinesis. They need a human agent in order to do so.

They can tap into one's pineal gland if the person is emotional and unaware of one's own vulnerability. The human pineal gland, at the very center of the brain, is the catalytic converter that will manipulate gravitational metaparticles into producing electrogravitic fluctuations. The person serving as the conduit for the entity is called the epicenter of the phenomenon. Paranormal activity does not come from the entities themselves, but from the subconscious – controlled commands done by the entities to their hapless victim's pineal gland.

The entity is a manipulator, and the victim is the one producing the paranormal activity, completely unaware of their own part.

The entity is not intelligent but sentient enough to allow it to have innate semi-intelligence. It will interface many neural memories of an individual to create an artificial identity, kept secret. It is kept secret because to know the true name of something gives one power over it.

Ironically, this is not a rule, but merely a belief that is implanted in the very psyche of human culture: the concept of an identity that is unique and unconquerable when unknown. So when tapping into a human neurosensorial memory-unit, these values are intrinsicly placed upon the "identity" created by the entity, therefore submitting the spirit to such reasoning, as well.

However, some believe that such thinking is a fail-safe cognitive primal mechanism to protect against the very idea of mind manipulation.

The big question is: who or what implanted such a psionic defense upon our psyche? I, myself, believe we are products of an intelligent design, and not just a random accident from a "nutrient soup" or genetic pool, caused by a meteor dissemination of panspermia. I do believe in evolution, but an intelligently designed one.

So, next time you think you are "only" human, think better, because, believe it or not, you are far more than that. You - we - are mighty humans. Even those pesky entities know that much. It is unfortunate that just a few of us know the extent of our strength. Knowledge is indeed power. Know more about yourself and your environment, and your views of both will drastically change. Regardless of what one might think, a great paradigm shift is on its way. When will it arrive?

CHAPTER 11

A REVERSE VIEW OF
AN OUIJA SÉANCE

From the view of the alleged spiritual realm, the view of a Ouija séance is quite different. While we cannot truly know, the following is a plausible conjecture of how a Ouija board séance would proceed, from the other side's point of view. The means used to obtain the following hypothetical account shall remain undisclosed as agreed upon sealed terms related to national security. The author will vehemently deny any knowledge of Project Mnemosyne's Operations.

The early method for exploring time and space was known as the Remote Viewing Process (R.V.P.), as used in the defunct Project Stargate led by the CIA. This account refers to the new generation of the Remote Viewing Process, known as Mnemosyne, and the doers (or the viewers), the mnemonauts. Mnemosyne is an expansion of one's own consciousness, reaching out through time and space. Mnemosyne came from the arcane Out-of-Body Experience (O.B.E.) during which a person has the illusory perception of leaving their own body.In reality, during these so-called "O.B.E."s, an ethereal probe is sent via a human-generated thought form. This produces a replica of the person experiencing the "O.B.E.," which creates the illusion of leaving one's own body.

The remote viewers use direct consciousness without a need for thought forms, whereas mnemonauts are cavaliers who must go beyond the space time continuum, without the anchors keeping them grounded to our common existence.

The following is allegedly from the personal notes of a mnemonaut participating in Project Mnemosyne. The author will not confirm nor deny this statement. As you will see, the mnemonaut encounters both an entity, named NOXEF-26B, as well as human-like creatures from another universe. The mnemonaut transcribes language from both parties with the help of Artificial Intelligence (A.I.).

Ouija Séance file – Operation #CLASSIFIED Session #CLASSIFIED

Contact established from adimensional atemporal non space (deemed as unconfirmed). After a considerable waiting/no time parameter in non-space, mnemonaut begins to describe the setting.

...and the background is an impossible shade of absolute white. It isn't illuminated; it is just plain white. It's whiter than any paper ever made. It is disorienting because there is no up or down, left or right. It is just this... whiteness going on forever. Definitely there is nothing here. It isn't like a vacuum because even in a vacuum there is always something. There is nothing here but me, to be precise there is not even gravity, light or anything to be measured, as a point of reference, at least for the standards of measurement that we currently use. No radiation is detected from nowhere. Temperature is... there is no temperature. No cold nor heat. It is just nothing. No gravimetric signatures as well... if there ever was a dead zone, this non-space should be it.

I know you can't respond, Control. But I will still continue to transmit. The only activity detected is my own and the echo of my point of entry, thanks to the beacon that I have left there. There is my only point of reference on this maddening place.

A target just appeared from nowhere in my sensors – elaborating.

...Readings are unclear, but it is definitely there. For some reason it seemed not to be aware of my presence, something that makes no sense, since there is nothing here but both of us.

Sent a passive psi-probe and trying to interlink. I shall run it through the psionic-buffer to isolate any traces of my consciousness. Interface would be done in... it's no use, time becomes irrelevant here... there is no means of counting time. I shall use my heartbeat pace as a measure of beatings of passage through time stream. Exchanging systems. Neuroservers accept new bearing marks and will use them to provide time equivalency. Mostly for my own sake of bearings.

It is one of the energo-parasitic life forms. Readings on quasimolecular biostructure point, a form of unknown radiant energy. No comparative parameters from prior encounters with other similar energo-parasitic life forms. I shall register this new finding and make a systematic list of observable readings with descriptive details. Table finding is NOXEF-26B (non-organic extradimensional [quasi] electron-based foreigner [bioenergetic lifeform], entry file 26, subsection "B").

Interface was done using a passive stream of neutrinos. All sources come from me. There is no neutrino activity on this side (or any other recognizable elemental particles). Elapsing linking is nominal. The entity is still unaware of the neutrino interface since they are uncharged elementary particles and are massless in this environment. I shall be receiving its metaconsiousness impulses in moments.

Received its impulses. The added logistic logo linguistic algorithm shall interpret them to their approximated equivalent on English language. The A.I. has to elaborate certain impulses with extrapolative interpretational language. It will not be exactly what the entity is thinking upon its consciousness, but it will be approximately enough for understanding of content. Transliteration enabled. Text follows:

"F-faint. Close to... now here. Track... signal/light/ source (?). Where... finding... to? No now here. Signal/light/source... near".

Would the entity register my presence, afterall? Seems unlikely.

111

It seemed to be referring to something else. Still, I shall remain prepared to instantly withdraw to physical space at the speed of one thought – continue with transliteration.

"*Need. Motive (?) doorway/entrance/portal/node (?) to realm of <untranslatable> located. Fix on... pathway/trail/road (?) to it. Calling... from... worthy... game/prey (?) listen to calling. To find game/prey. Calling from game/ prey I listen. Solids (?)... solid/material (?) game/prey. Good/acceptable source nourishment. Good/acceptable source hungry good/acceptable. Feed (will) I. Find (exact) pathway/trail/road (?) entry point. Entry point. Fix upon. Fix upon entry point. Access game/prey (?). <untranslatable> located. Well (source?) of game/ prey (?) call view/know/(unseen) sight. game/prey (?) <untranslatable> located. Pathway/trail/road (?) fix upon. Conjoin/unite/connect*".

Control, I am receiving the signals from the NOXEF-26B. I retrieved signals via some transdimensional conduit. I can't find any readings to pinpoint a location, but the entity certainly can do that. A.I. informs that it can recreate a visual equivalent of the signals the entity is receiving by adding a graphic-inducing algorithm to the code lines. Processing sequences. Describing the visual input: the code lines. Processing sequences. Describing the visual input: no way! Is that game! The spirit's thing. Whatchamacallit... oh! Ouija? No, no. It's the Ouija thing. A.I. is cross-referencing metadata's vault. It is a transdimensional version of the Ouija board, transliteration points out a dimension very similar to Earth's own. A.I. working on the native language of this Earth-version. Not exactly English, but some sort of Slavic-based similar language. Interesting. Linguistic interface completed. Translation to English language shall proceed from native dialect – the people look very similar to humans. Still not able to access biological interfaces for specific biological readings, all data received is coming from the neuromatrix of the entity itself. These people could be almost exactly as humans from our earth. The fashion style is a mix of distinctive human ethnicities. Not able to set equivalent bearings. Yet the furniture evokes some twentieth-century human history. A.I. extrapolates this is the equivalent of our 1960 or so. It is a group of teenagers between apparent ages of 14 and 16 y.o. give or take. Three males, two females. The process of us and the board marking seemed eerily similar to our own Ouija board. Back to the entity's neurofeed.

"F-fresh/edible-hunger/hungry. Good well <untranslatable> Key. Bright. Key. Way. Enter. I. Push (?) key. Enter."

Control, something is happening. Readings point to something akin to gravimetric fluctuations, but the A.I. confirms that they are not graviton-based. Some sort of shimmering bearing coordinates CLASSIFIED. The area is "tremulous." In the absence of a better word to define it the site started to shine waveringly... something like sparkling or glimmering but not quite that. Get my neuro-optical transmission because I lack the proper words to describe what I am seeing. Entity's neuro feed resumes.

"Hunger/hungry. Need. Key. Brighter (?) way. In. Enter. I".

Seems it is approaching the site and... wow!!! An opening! It is like a stable quantum singularity, but without any recognizable gravimetric readings! I can see the other side! It's physical space! The damn thing opened some sort of stable wormhole to a physical dimensional universe like the very same one we saw on its optical neurofeed.

Readings gave a measuring of 98.2% of similitude with our earth. In fact it is on the records. A.I. points it to be universe CLASSIFIED, the same one of Dr. CLASSIFIED's team discovered last year. Another Earth twin. Seventy-two and counting! There is indeed a multiverse, people of little faith. Sorry, Control. It's just way too exciting!

Getting audio feed from site through the opening. The entity still is not crossing. It seemed to be studying (or whatever) these kids. Why doesn't it do anything? Audio feed enabled. From universe CLASSIFIED, audiofeed of natives:

"...are there? Please, don't laugh, Mar'ahood. This is serious".
"D'oma-lah, shades (?), really? Friends tell me if D'omma- lah is not unwell... to try to summon shades... using a D'ja-lah board?".

"Mar'ahood, please, keep your skepticism to yourself. D'oma-lah graciously invites us to her domicile to engage in experiments using the D'ja-lah board as you well know. Why did you come if you felt this to be nonsensical?"

"Venerable Tiur-ge, I know you are the daughter of Pak'ein the Seer. I praise your progenitor's noble trade. But I still do not comprehend how a piece of <difficult to translate, but would be something like "pulp of wood, or something similar"> would be the venue to contact a elsewhere?".

"Mar'ahood, the skeptic. You would not even believe in (rain) even if you were getting wet with it. I am a true son of Verd'ak, the Counselor. My kin accept the abilities conveyed by the D'ja-lah board. Be skeptic if you wish but do not insult our host."

"Much obliged, noble D'enok, true son of Verd'ak. I am grateful for your defense. I do respect someone so young being wise and consonant as you are."

"I humbly declare to be a son of my father, fair D'oma-lah."

"Oh, for A'kron's sake! Cease such unnecessary evident mutual courtship. We all know how you two are keen to each other. But, as always, noble D'enok has a valid point. I humbly beg your forgiveness for my blunt words, fair D'oma-lah. Would you place your forgiveness upon me?"

"Of course, Mar'ahood, it is given. It will be my most humble pleasure to have you, Tiurge, D'enok, and Beernah'e on my domicile. Since our progenitors shall be returning soon from their mutual gathering, should we not continue with the shades summoning by the D'ja-lah board?".

"I speak for all, fair D'oma-lah. Proceed by all means."

"Much obliged, Mar'ahood. Please, gather in a circumference, and place one digit upon the D'ja-lah board. Concentrate upon the Shades's realm. Beg for a link to it."

Control, are you listening to the audio feed? There are some weird kids! Did you ever hear kids speaking like that? It might look like Earth, but these kids seem more like aliens than E.T. They have no clue they are being observed by the entity and me. The NOXEF-26B is still inactive, doing whatever it is doing the audio feed refuses. From universe CLASSIFIED, audio feed of natives:

"... voice. Again, Shades of the after living, pay heed to my humble voice. I D'ama-lah, daughter of Tis'okh, the Trader, in company of noble D'enok, true son of Verd'ak, the Counselor, venerable Tiur-ge, son of Pak'ein, the

Seer; fair Beerlor; venerable Tiur-ge'gedo, son of Sun'tar, the Learner; and the Wonderer, Mar'ahood, son of Jos'aab, also a Trader, ask for your presenting upon our gathering, bringing word from the realm of the after living. We call upon you though our D'ja-lah oracle."

Control, the entity is showing signs of activity. A neutrino-like tendril is entering universe CLASSIFIED! It entered! The tendril is approaching the kid with a brighter aura. Did you receive the video feed from the spectrometer, no? All the five kids have three strong auras surrounding them. Not so unlike what we, humans, from Earth have. But theirs are way brighter, somehow. Anyway, I just wanted to make sure you are gathering all the records. This is amazing!

Well, the tendril is touching the kid named Tiur-ge. Seems like the kid felt something because he started to shake.

"Venerable Tiur-ge! What is happening, are you unwell?"
"N-no, fair D'oma-lah. But something is here, I can feel it."
"A shade, perhaps?"

Control, the tendril detached from the kid. It seemed to be searching for a specific person. Here it goes again. Now it is touching the girl named Beernah'e. She seems to be the youngest of the group. It is encircling her, but she seems unaware of this. The tendril merged within the girl's aura. She still seems unaware of that.

A bluish glowing is coming from the girl's aura. It is going through her right hand, the one touching that Ouija-like board thing.

The board...Control! It is enveloped by the bluish aura coming from the girl.

The planchette-like thing is moving by itself showing whatever letters are those in the strange glyph. Like the alphabet of these people. It seems to be spelling a message. I have no clue what it says because I surely can't read their language. Audio feed resumes.

115

"...ing! It moves! By A'kron's sake, you were right, fair D'omalah! The D'ja-lah board is indeed a pathway to the Shades!"

"Of course, Mar'ahood. I wouldn't have you all to waste your time on the account of silliness. Shade of the after-living, let us know that you are indeed here. Present a signal of factual proof."

Control, a tendril of bluish light is touching the nearby wall. It is coming from Beernah'e's head, it seems to be pulsing. Registering gravimetric activity. Kinetic momentum is being produced by acceleration of molecules. It is a sort of directing gravity field.

"It knocks in the wall! Indeed, it proves its ethereal (?) presence over physical matter. Praise A'kron! It can hear and understand us. Who are you, Shade? Identify yourself, please."

Control, I can detect tendrils going through all the kids' heads now. The entity is not absorbing lifeforce. It is data. Mnemo-memories, I believe. It is reading their subconscious memories. It is learning from the kid's memories.

The electrodes blink upon the Ouija-like board. It is not just made of wood as I've thought. Readings showed a crude basic circuitry with LED-like mini lamps embedded within the glyphs of the board. The planchette-like thing touches a symbol and the LED-like lamps light up a purple light. It's weird, but it is also neat, somehow. Why have they never made something like that with our Ouija boards back home? I bet it would sell as well as jelly doughnuts. Resuming audio feed.

"Mord'ach? It spelled / wrote the name Mord'ach. The Mord'ach? The glorious leader of the Kiappe'alal order?"

"But, D'emok, he died ninety (?) gargja (years?) ago! It is not possible. The Kiappe'alal order does not believe the Shades to be real. They are technopriests. They believe in the Cyberafter living."

"And yet, Mar'a hood, the proof lays at your very sight. Have you not been humbled enough today, Mar'a hood?"

"I stand corrected, noble D'emok. I do see such proof in sight."

"Mord'ach, I, D'oma·lah, daughter of Tis'okh, the Trader, beg for your

word. Which message do you have for your summons?"

Oh-oh. Something is going on, Control. The tendril is becoming red now. The girl Beernah'e, started to tremble. Something was triggered here. Her entire aura became red. She is speaking with a guttural voice. Seems like a male's voice. Resuming audio.

"Infidel! How dare you follow the heretics of the Kiappe'ahal Order?! The fiends murdered me. They are enemies of A'kron! Oh, I do know your secrets, vermin. You hide from others that you seek knowledge of the Kiappe'alal order, Mar'ahood!"

"No, do not listen to this phantom. I despise the Kiappe'ahal order as you all do. These are untruths said by this ignoble shade!"

"Are they, vermin? Take his information processing unit (a laptop?) and see the truth of my denouncing of this infidel."

"B-but he is the shade of Mord'ach! He once led the Kiappe'ahal order himself. How would he turn against his own kin?"

"Silence, Mar'ahood! Is this true? Are you a follower of the vile Kiappe'alah order? Are you a believer in the abominations of the cyber-afterliving?"

"Fair D'oma-lah, I am not a heretic. It is true that I have looked upon the forbidden feeds on my information processing unit."

Control, do these people have laptops in the 1960's? Shoot! Imagine what they would have nowadays? Resume audio.

"Again, I repeat, Mar'ahood: Are you a Kiappe'ahal follower?"

"I – I do not know, fair dame. I was...am ... curious about their ideas. I never met none of them, but I indeed looked upon their knowledge."

"Liar! Deleting the messages from the one named NAA'GEER will not save you from your sins, cowering vermin!"

"H-How do you know of NAA'GEER?! No one but I and him know of our palavering. You are indeed a Shade talking via the hapless Beernah'e."

Control, the A.I. said that the biosignatures of all sorts are running in flowing pathways. The cross-referencing matches as an "emotional storm." Seems like most emotions such as fear, rage, disgust,

117

and dread generate quite a flow of biofeedback energies from these kids.

NOXEF-26B is gorging on those kids. It is draining them kids dry.

I couldn't interfere with it even if I had the means to do so.

You know what the entity would do to me if it knew I was here, no? I'm sorry for those kids but better them than me. Besides, they look just like humans, anyway, the lads are alien and ... ok, ok. Remove this part from the records. No, I'm not xenophobic! But this might look like Earth, but it isn't Earth, and I'm not here, capisce? Let me do my job and get back to our universe before I become the next meal of this nasty one. The damn thing duplicated in size. And... oh - oh... Crazy! It... I don't know... split a part of itself... a smaller part. It just split itself in two or just gave birth to something... whatever... that smaller part of it has the same biosignatures of its... larger part... or mother... or whatever. The smaller... whatsoever... entered dimension CLASSIFIED. The portal/entrance closed. The big thing stayed here, the little thing left to... the place with the kids, listen, I will make a last neurofeed on them, then I'm out of here. I don't want this thing to eat me too, ok? There it goes.

"...well/good. Satisfied. Repose now. Seek more. After. Let open key after. Gather extension of self-back. It be satisfied, too, then. Go make sources (?) send more sustenance with... feelings (?) like visit sources. Good having key. Repose now..."

Well, that's it, Control. I'm out of here. The thing just blinked out of existence. Just like that. I think it's still here, but I can't sense it anymore. NOXEF-26B, nasty as all the others.

Will be debriefing after recovering from my return. Mnemonaut CLASSIFIED signing off.

Ouija Séance File - Operations #CLASSIFIED Session #CLASSIFIED Abridged Version of Briefing (Edited Version) - After the Debriefing

After reviewing all the data provided by the mnemonaut #CLASSIFIED, the head of the department declared the following as the outcome of the debriefing:

REPRESENTATION OF
BEERNAH'E

As observed, another NOXEF was found and classified under a distinct category. The non-space sector visited this time had the same confusing and amazing sterile characteristics as the other sectors mapped. The dead zone, the non-space, is still being mapped by the mnemonauts sent within it.

The finding of the NOXEF's was a surprise to all of us. As they have an apparent inability to perceive the mnemonauts, even at a reasonable closeness to them. I'll explain some of our tech to the Pentagon's envoys.

To allow mnemonauts access to instrumental gear essential to their missions, ectoplasmic[1] replications of functional solid parts and devices using the protoplasm[2] of the very mnemonaut are produced. The side effects of the medium-term exposure of gear recreated at nanoscale via irradiations of element 118 (Oganesson, formerly known as Ununoctium) would cause their counterparts to be replicated to ectoplasm and controlled by the mnemonat's willpower. An Artificial Intelligence program was inserted into the components of the gear, including apparatuses of beacons, psionic probes, and transdimensional telecommunications arrays.

NOTES

[1] Ectoplasm – The ethereal counterpart of the protoplasm. It is capable of reproducing the "echo" of an image of any device built with protoplasm as the source of engineered structure of biological mass. Ectoplasm is resilient and extremely malleable. Although ethereal in essence, it has its own gravity field that allows its replicated parts to become functional, making it the perfect material to be used to build devices for use in atemporal/adimensional non-space. It disintegrates by itself after some time.

[2] Protoplasm – The complex colloidal submicroscopic particles made largely of living plant and/or animal cells. In this case, they are used in nanotechnology to recreate robotic and instrumental devices at nanoscale.

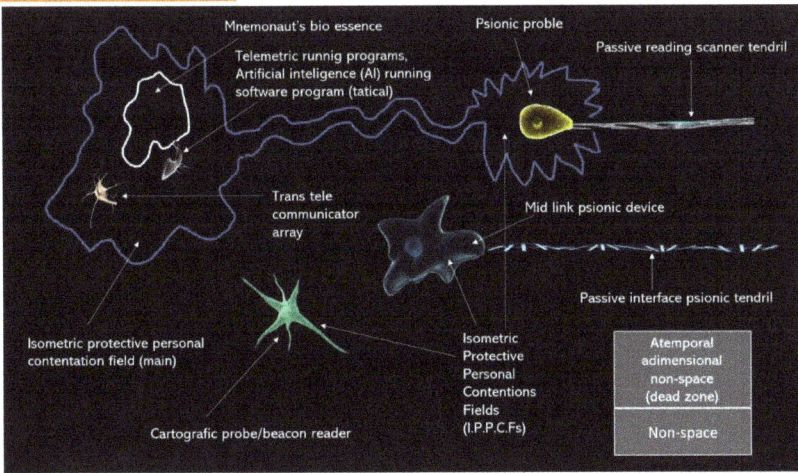

Mnemonaut's bio essence
Telemetric runnig programs, Artificial inteligence (AI) running software program (tatical)
Psionic proble
Passive reading scanner tendril
Trans tele communicator array
Mid link psionic device
Passive interface psionic tendril
Isometric protective personal contention field (main)
Isometric Protective Personal Contentions Fields (I.P.P.C.Fs)
Atemporal adimensional non-space (dead zone)
Non-space
Cartografic probe/beacon reader

To the benefit of our Pentagon guests, this is the breakdown of a typical mnemonautic incursion on atemporal, adimensional non-space.

- The Isometric Project Personal Contention Fields (I.P.P.C.Fs) are fundamental in providing protection of the mnemonaut and their gear against the Dead Zone. The essence of the non-space is devoid of any known energies, so we came up with a form of variation, called wave multi-polarizing. A shifting spectrum band resonator system isolates the essence of the mnemonaut and their gear from harmful direct exposure to non-space's environment (or rather, non-environment). Its presence in the Dead Zone should be no more than sixty minutes, or the fields will start to lose integrity. As seen on prior exploratory probes, the effects of direct exposition to non-space are anything but desirable. We keep our sojourns at most fifty minutes, leaving five extra minutes for any emergencies and a built-in, completely automatic system of retrieval and exit for immediate return that removes the mnemonaut and its gear at the start of 59:00 minutes from the one hour deadline. We have lost probes, but a single mnemonaut has never been harmed or lost.

- The Artificial Intelligence (A.I.) acts as the mnemonaut's tactical advisor, consultant, and co-pilot. It acts as communications officer, engineering advisor, and psychological support as the mnemonaut is venturing into absolutely unknown territory. If a mnemonaut were to lose their consciousness, the A.I. would take over the mission and will exit and return immediately.
A constant 360° probing is done by the A.I. that will advise the mnemonaut of any significant happenings and/ or intriguing elements that require human attention or supervision.

- The Transtelecommunicator Array (T.COM-A) is another vital piece of gear and the first successful piece of equipment to use the Quantum Sub-particle Transdimensional Communications System (QSTC-S). With its use, it is possible to maintain contact with the mnemonauts in almost real time. Variations of seconds or nanoseconds may occur depending on the conditions of the environment. The transmissions originate from dimensional sites in extremely short time (femtoseconds), transdimensional sites (picoseconds), and the distinct regions of dimensional atemporal non-space (nanoseconds to seconds depending of the point of origin of the transmission). Replication of the ectoplasmic engineering from protoplasmic constructs was tricky and took about a twelve-year period to be properly achieved. The basic fundamentals are: create a digital 3-D model, build it on a Quantum molecular 3-D printer, recreate a functional version of it using colloidal proteins (hybridized plant and animal cells) of which each and every part made of protein will produce the same function of a device made of metal or synthetic material. Once the colloidal protein device is done and proves functional, it will go through a further process, which is classified.
An ectoplasmic copy of this device is created, but due to the Reflexing Echo Effect, the ectoplasm loses integrity after one

hour, hence why no mnemonaut whatsoever, can remain in action for more than 59 minutes with fifty-nine seconds to spare.

- The Telemetric Running Program (T.R.P.) is the data-gathering of measurements made by automatic instruments (and overseen by the A.I.) scanning from a distance within the 360° radius of the I.P.P.C.F. main station. Range is variable but usually approximated to be one mile in human measurement. The T.R.P. is fundamental for the cartography of the distinct explored regions of transdimensional universes, and the challenging new expectations of non-space territories. So far, less than a fraction of both nodes have been mapped.

- The Cartographic Probe/Beacon (C.P.B.) serves two important functions:

  o To do individual-proximity readings of specific chosen sites.

  o Act as a beacon for bearing frequency coordinates. Distinct from the other ectoplasmic devices, its metamolecular structure is made to ensure it will leave a residual tracing of its last position even after it is dissolved by entropy, making it a permanent marker used to provide bearings on a site that has no other landmarks to provide directions.

It was built to be an autonomous independent unit that will transmit data and receive instructions via the mnemonaut's I.P.P.C.F. main station. It has its own I.P.P.C.F. unit to protect it from any harm from non-space. It also has autonomical displacement capabilities.

- The psionic probe (PSI-P) was not created by our research & development programs. Its design and specifications were retrieved by a mnemonaut's expedition searching for new technologies on other universes, specifically in other versions of our Earth through the multiverse. It was discovered in a metahuman civilization over dimension CLASSIFIED, after the defunct psionic Project Stargate (from the Department of Defense under the given authority of the U.S. Air Force) and it was done on mnemonautics pre-era by the now-obsolete Remote Viewer's Program. But, nevertheless, the Remote Viewers created the foundational elements that would later become key elements of mnemonautics. Project Mnemosyne was only possible because of Project Stargate.

The psionic probe is a device that links to the consciousness (or even proto-consciousness) of beings and entities to access their mind-matrix thought-process system's configuration.

It is done in three-phases:

1. Interface one - seeks intent by multi frequencies of emotional state;
2. Interface two - seeks comparative patterns of linguistics based on algorithms of cognitive extrapolation;
3. Interface three - establishes a basic translation program for thought-expressions, translating it to English language.

The entire process is done passively using a neutrino field as the medium.

- Mind-Link Psionic Device (M.L.P.D) allows the mnemonaut to not only receive the thoughts of the being or entity but also to see, hear, and feel whatever sensorial input the being or entity is experiencing themselves. It can turn these neurosignals into coding to record both audio and video of all that is experienced by the target of the mind-link psionic device.

Like the psionic probe, the M.L.P.D. has self-sustaining independent drives, but its creation wasn't transdimensional, but extraterritorial, in origin.

The technology and knowledge acquired with both initiatives, the defunct Project Stargate and the contemporary Project Mnemosyne, changed and evolved our capabilities in ways we could never have dreamed of. Indeed, we have gained a lot on the acquisition of extraterrestrial technologies, but it originated in our own universe, and therefore, it will fall under most known dimensional elemental capabilities. Still, having acquired not only technology and knowledge from other universes, but also exceptionally unequal sources of power, has put us humans in a position where even the outerworlders must understand we are a force to be reckoned with.

The Ouija board is seen merely as a toy by many. If only they knew how useful a tool it can be to pinpoint transuniversal gateways, they would be shocked.

Although our own universe is sometimes a victim of these ethereal predators, we do not yet have the means to avoid their predatory action against hapless humans. Furthermore, some reason that we need these very entities to both find and access other universes that are still out of our willing reach.

Most experiences with the Ouija board end in cheap scares and thrills of no greater consequence, yet one should always be wary of dangerous, or even fatal consequences. We mourn those we have lost, and honor what we have learned in their memory.

Gentlemen from the Pentagon, we understand perfectly your dilemma and are truly grateful for your support in all the distinct areas aside from our granted budget. If it serves any comfort, know that even if we were prepared to give combat to such entities, there is no way to patrol the entire non-space.

We know quite well that we cannot turn these issues public. What would the people think of their tax dollars being used for projects involving Ouija boards, and worst of all, Ouija boards from other universes? What do you think the constituents would do to us? Or to you, even? So it is best we keep this under wraps and keep doing what we do best: producing worthy results. We know that we can not keep secrets forever, as seen with the mess your own people did with the alien business. At least it is easier in our case. If the word ever gets out, who would believe it? Ouija and scientists? Spirits and the military? See the beauty of it? It's a win-win situation. Thank you for your presence and attention, gentlemen and ladies, and we give end to this briefing.

CHAPTER 12

A BASIC VIEW OF OUIJA'S NON-SPACE BIODIVERSITY

The underground Mnemosyne Initiative has a division for research and interdisciplinary subjects related to the biodiversity of adimensional atemporal non-space.

The profile of a researcher requires absolute confidentiality determined by a vetting process, availability to travel abroad and beyond, comfortability with mixed disciplines of research, and a willingness to straddle field boundaries. A researcher must be one who wishes to explore uncharted domains, is capable of creating collaborative sub-projects, and who thrives in an extremely unusual and even dangerous dynamic environment.

The Mnemosyne Institute for Integrative Biodiversity in non-space exploration is a world-leading center for such research.Its central mission promotes theory-driven synthesis and data-driven theory in integrative biodiversity research on a medium outside our own plane of existence.

The concept of this effort encompasses the detection of biodiversity, understanding of its emergence, exploring its consequences for environmental functions and capabilities, and developing strategies for the safeguards of the mnemonaut's and their equipment over non-space regions. Neuroscientists in this effort study altered dopamine and mutated neurocells of the mnemonauts.

The main task of this working team is to conduct independent, interdisciplinary research of non-space environment and related topics using empirical approaches, such as field observations or manipulations, mesocosm/lab experiments with available data and samples in order to provide analyses, and address possible evolutionary questions in a clear biodiversity context.

An investigation into the biophysics of developmental processes of ethereal entities (using our understanding of the entities' neuromatrix) would provide a better understanding of their behavior. It would also provide us insight into possible perilous responses should direct contact occur.

Dopamine is an organic compound that appears as a neurotransmitter in the human brain. It is the main key ingredient in the biochemical solution that helps the mnemonaut to achieve their neuro-inducing capabilities when interacting with the replicated ectoplasmic constructs. Such high-affinity responses depend on a competitive selection of neuron cells carrying somatically mutated neurocell receptors into key centers of the brain. The rapid neuron cell interactions that occur during the process of connecting to the ectoplasmic constructs also resemble neural synaptic transmission pathways.

Mathematical modelling computed that faster dopamine-induced somatically mutated neuro cell receptors increase the total input/output of neural centers and accelerate it by days. Delivery of neurotransmitters across the mnemonaut's synapses is certainly advantageous in the face of a non-space entity. It gives people a better chance of survival in a non-space environment.

The findings of these researchers gave us the following brief view of the non-space biostructural and elemental biodiversity.

Biostructural and Elemental Biodiversity

- Miasmas are made of compressed electrons that operate somewhat like radio waves and electromagnetic waves. Within the radio wave spectrum, miasmas would be akin to the longest wavelengths that range from the length of a football to longer than a football field.

128

- <u>NOXEFs</u> (Non Organic Extra-Dimensional [Quasi] Electron-Based Foreigners) are high-energy electromagnetic-like wavelengths shorter than those of ultraviolet waves. Similar to gamma rays, they have the shortest wavelengths and highest energy of electromagnetic waves. The conjunction of these metaparticles requires a cognitive factor in ways not yet understood.
- <u>Spectres</u> appear within the electromagnetic spectrum in similar patterns to infrared waves. Their wavelength is shorter than radio waves but longer than that of visible light. Spectres are capable of interacting with people and animals.
- <u>Apparitions</u> are residual effects of extreme psionic discharges produced by enormous emotional output. They appear in cycles, dependent on a trigger factor, such as emotional distress, and can portray themselves under the visible electromagnetic spectrum humans can see. These converted visible rays have patterns that are recognized by the human eye as color definition.
- <u>Phantoms</u> are similar to apparitions, but in this case, the triggering effect occurs when the light energy is both radiated and absorbed as tiny packets or bundles, and not in continuous waves such as the apparitions.
- <u>Bogeys</u> use the electromagnetic spectrum but require a gaseous medium, such as organic-base atmospheric substances like methane gas, for instance. They operate under higher frequencies than visible light waves. They are sentient but not cognitive and act on instinct. They feed from emotional output, namely fear, the simplest but most powerful emotional source.
- <u>Eidolons</u> operate in the short wave spectrum, with distinctive variable properties of ever-changing wavelengths that are visible to the human eye. This quasi-sentient entity lives on emotional wells of stasis, waiting for a trigger factor, just like apparitions and phantoms. Using the environment's residual energies as basic sources of nourishment, they wait for humans to get scared and produce a significant amount of bioenergy from their emotional discharge.

This research on non-space entities provided a significant structural engineering model, bringing analysis and more technical understanding on a subject that was formerly dismissed as "supernatural" and part of the so-called "occult".

An important part of this project was to provide technical analysis on the natural (and the not-so-natural) elements of such entities: to give us a better understanding of frequencies of motion. With an understanding of these surreal but distinct frequencies, we may be able to identify and avoid dangerous confrontations with outerworldly entities.

Interferometers are essential in scanning interference of wavelengths for precise measurements. We have learned that as a wavelength's frequency decreases, the length increases. When electrons in subparticles interact to form such wavelengths, bonding creates a stored energy of electromagnetic force fields, that make up their fundamental structural foundation.

These seven basic examples of the non-space biodiversity also provided insight into how they interact within physical dimensional space.

We know that we are still far from fully understanding these entities, but as was their intention, this elusive and secret black budget project's team did establish a foundation of knowledge on non-space biodiversity.

Next time you decide to use the Ouija, you will know that it is much more than a simple game. The who, when, and where will determine if that session is just for leisure or if that session allowed something far more insidious and dangerous to enter our dimensional world from the realm of non-space.

You don't need to believe any of this. A shark will find a bleeding person in its waters, regardless of whether or not the person believes themself to be a target. Independently of belief, the shark will find you if it is around, and so will the spirits from the realm of the Ouija. When you finally realize or begin to believe something is really there, it may, perhaps, already be too late.

EPILOGUE

One of the most persistent beliefs of humankind is the idea of ghosts, a spiritual essence of the dead that continues after one's demise.

I am a firm believer that in the universe nothing is truly created nor destroyed, but merely transformed. I do believe we are essentially eternal and will continue to exist after death.

I am far from having all the answers and would be an idiot or a deceiver if I claimed I did. But all that I have come to know has made me a firm believer in the survival of the human soul.

The tradition of the Ouija is based on claims of communicating with the spirit of the dead. According to testimonies, spirits had to be who they claimed to be because they possessed specific knowledge only those involved would know. The mythology of necromancy, the arts related to the dead, was born by such beliefs as well. The Ouija board simply became the more popular venue for communicating with the departed.

All of these traditions are based on belief, based on faith.

Faith can be defined as a mostly unconditional reliance or trust in someone or something. Faith is trust, is a belief, a confidence, a conviction, a credence, a reliance on acceptance. Faith is not only good but also necessary. It may bring balance to one's life.

Blind faith, however, is senseless, and may cloak the true essence of someone or something.

Ouija practitioners believe in the board. They have faith in its workings, even if they are also blinded by its mechanics (though most would never admit that openly). No one wants to address the truth that the Ouija is not a toy to be played with.

The Ouija is indeed an instrument of communication, but not with the dead. The Ouija is a window for communication with transdimensional parasitic ethereal predators, which feed from one's emotions and even one's own lifeforce. We must expose these entities for what they are: sentient cognitive intelligent parasitic predators that lie, deceive, and cheat any prey to satisfy their needs.

It is so much simpler to accept the ghost of a departed one than a horror from another dimension preying upon someone's faith. Be wary, for prey does not need to know they are prey for the predators to prey nevertheless.

Modern spiritualism must evolve to guide the new brave minds of the twenty-first century to confront today's hocus-pocus that follows the same line of mystifying principles from eras past.

We need a new spiritual renaissance that will lead us to where we should actually be.

So, who knows about such entities? Even if the common public does not know, certainly the figures in positions of power would've had to be briefed on a menace of such magnitude. Notice that I said "figures in position of power," not the government. Conspiracy theory? Yes, there is indeed a conspiracy of silence within the hidden echelons of power about this and many others deemed "sensitive subjects."

But it's no theory. It is entirely factual, unfortunately. There are no evil governments conspiring against their citizens, but simply people in positions of power using information they believe to be rightfully theirs, to maintain their balance and power.

The metaphysical divisions of such secret echelons study all anomalies they see as relevant. They have knowledge, and with that, comes power.

Needless to say, these hidden powers have known about such entities for a very long time. Their objective might be to understand them so they can control or destroy them. Their clandestine laboratories send out all sorts of information-seeking probes to gather knowledge. Such probing revolutionized the way these underground scientists see reality as we know it. Once explored, scientists found things were not as they seemed to be.

Ultratelemetric probe readings mapped already larger structural ethereal constructs all over non-space. The utmost goal was to probe without being detected, so that if the probing did cause any disturbances, they would be so minute as to be unnoticeable by the entity.

The Ouija is nothing but a bridge between our realm and the adimensional realm of these entities.

But if they are not ghosts of the departed... How do they know so much about their lives, about the people involved, and about the circumstances? Only the dead would know that, no?

Not necessarily. The entities are intelligent enough to know they need an emotional hook in order to be grounded to a person or site. Such entities can access your memories and obtain all the information they need in order to convince someone that they are indeed the dearly sought departed. It is deceiving. The entity will feed on the emotional output of such interaction, be that joy, fear, anger, or whatever other strong emotion that may surface.

Because of my research, I have learned things that I wish I had not. I discovered issues that made me wonder about our civilization and even about myself. Indeed, I've seen wonders but also horrors, as well. This made me fear certain humans more than any outerworlders.

So many times, I had thought of giving up such things but exiting the cabal is even more difficult than entering it.

After my end of term with the Blue Planet Project, I went to another, the Pulsar Project. My last term was with the Nemesis Project. I was caught in a web of underground black budget projects. Nobody forced me to be part of them. For that I have no one to blame but myself.

What I learned there about outerworlder subjects and advanced metaphysics turned out to be more than a worthy price to pay for them. I won't go into details about my hiatus and how I was away for decades. I intend to write a book about this, and it will be a thing that will shock and amaze you.

The Blue Planet Project was never meant to be published as it was. It was published by others, taken from personal notes in my field journals. It is full of anachronisms, in addition to being ciphered, so its real knowledge will be known only by me. It was the only way to keep my notes safe in case of interception. At first view, the unsuspecting observer will see the anachronisms and downright inconsistencies and will not give any credit to the important knowledge hidden there. In the end it was better this way. Let the opportunists do what they must with the books. They have no idea what they are selling there.

On a personal note, such scoundrels are taking advantage of my intellectual property, since I never earned one penny for such books.

Now, my own books are to be available soon, this being the first of many yet to come. No more codes; everything will be out in the open.

Why now after all these years?

I suppose I finally learned that life is way too short and, when you are down, there is no other way but up. If not I, who? If not now, when?

It will be David versus Goliath, all over again. This may turn into a very interesting and unusual situation. And I'll do what I can in hopes of a happy ending.

Get ready for a new line of books like no other.

Know the distinction between learning and knowledge.
Knowledge is borrowed, learning is yours.
Knowledge is acquired with the use of concepts.
But learning never ends. It is always ongoing.
A seed is an introversion of learning. It is a centripetal force that moves inwards, toward the core of our very selves.
This is why a seed is completely covered and closed to the world. It is lonely and solitary without roots in the earth or links in the sky.
A seed is an island completely isolated from everything.
It is unrelated, having no doors or windows to the world.
We are all born with seeds within ourselves.
Another name for them is souls.
That is what the entities seek from us.
To want is not enough to give birth to this seed.
Only the flow of <u>true will</u> *can do that.*
In a sense, you are the obstacle standing in the way of balance and happiness
Remember, books like this are here to give you options.
But in the end, only you will decide if you have learned something from this book or merely accumulated knowledge.
I can offer you guidance, but only you can reach a destination.
HAPPY TRAILS!
Think about this small book as an <u>AWAKENING</u>.
The choice of a rebirth into learning is yours to make – and favorably to take.
Rebirth: The gift of learning is a gift of new life.
The mind invites to nothing as the original nothing once was. Make this nothing, something.
It will make learning the feast of all feasts.

Peace and Joy!

-Jeferson de Souza
Paranet Global
Earthstation Seven
Georgia, USA 2018

FINAL DISCLAIMER

Any references seen in this book regarding the terms Mnemonautics, Mnemonaut or Mind Extension, regarding any military agency whatsoever, is to be disregarded. The author will not confirm nor deny any such subjects.

REFERENCES

The Encyclopedia of Ghosts, Cohen, Daniel New York: Dorset/1984

Real Life X-files, Lexington: University Press of Kentucky/2001

Physics and Psychics: In Search of a World Beyond The Senses, Stenger, Victor J./1990

Future Science: Life Energies and the Physics of Paranormal Phenomena, Garden City, NY: By White, John, and Stanley Krippner, Anchor Books (1977)

Haunted Places: The National Directory, New York: Penguin Press, Hauck, Dennis William (1996)

The Encyclopedia of Ghosts and Spirits 2nd Ed. New York Guiley, Rosemary Ellen: Checkmark Books (2000)

The Encyclopedia of Religion Vol. 5. New York: Macmillan, Eliade, Mircea (1995)

Handbook of Unusual Natural Phenomena, New York: Grammercy, Corliss, William R. (1995)

When Ghosts Speak: Understanding the World of Earthbound Spirits, New York: Grand Central (2007)

ESP, Hauntings, and Poltergeists: A Parapsychologist's Handbook, New York: Warner Books Auerbach, Loyd (1986)

Investigating the Paranormal, New York: Helix, Cornell, Tony (2002)

How to Hunt Ghosts, New York: Simon & Schuster, Warren, Joshua P. (2003)

In a Dark Place: The Story of a True Haunting, New York: Villard Books (1992),

Warren, Ed, Lorraine Warren, Al Snedeker, and Carmen Snedeker

The Haunted: One Family's Nightmare, New York, St. Martin's (1988) Curran, Robert, with Jack Smuri, Janet Smuri, Ed. Warren, and Lorraine Warren

Appearances of the Dead: A Cultural History of Ghosts, Ambeest, NY: Prometheus Books, Finucane R. C. (1984)

Evaluation of the Military's Twenty – Year program on Psychic Spying, Skeptical Inquirer 20, no. 2 (March/April) (pages 21-26) by Human Ray (1996)

Apparitions – London: Society for Psychical Research, Tyrrell, G.N.M. (1973)

ABOUT THE AUTHOR

Jeferson Borba de Souza was born in the south of Brazil, in the city of Porto Alegre, capital of the state of Rio Grande do Sul. The 3rd child of four, he encountered the uncanny at the young age of seven by the means of close encounters of the fourth kind, mental interaction with what he thought at the time as "beings of light." Too young to understand the reference to alien life forms, he confounded them with "angels." The beings told him that he and a few others elsewhere were selected long ago through a lineage of families. The purpose of such connection was to supervise certain members of these families that would become unique individuals with peculiar abilities. The amazing thing was that the children they reached for, Jeferson included, somehow understood all these beings told them, things that common children of such age certainly would not have been able to comprehend.

Jeferson learned the need for secrecy and kept his mind encounters (communications started initially in one lucid dream and continued in the form of telepathic interaction) to himself. It is important to say that the telepathy used by them was NOT the telepathic concept usually known. It wasn't reading of the mind or talking to the mind but an interchange of images and concepts that transcends communication as we know it. It was like the subject's minds became a collective one and even the basic senses as smell, sight, hearing, taste and feel were exchanged between the connected minds. His childhood was spent learning the plethora of knowledge of the available database allowed to the children they reached .

Jeferson's childhood was lonely as he grew socially isolated from his brother and sisters, navigating the world he lived in and the secret world that no others could know about, yet. Never having true friends and unable to share his secrets with his kin growing up was a challenge that sometimes seemed almost unbearable. His early teenage years let him know that the beings communicating with him were starfarers from the Vega star system, at the Lyra constellation. A main world, three colonies, two mining and three support worlds made up the Vega star system, which is surrounded by a large ring of planets and asteroids. The beings presented themselves as humanoids looking very much like humans from the country of India, with darkish skin and fair hair with Caucasian features. But later on, much of these would change as Jeferson grew into a young adult.

In his teenage years, Jeferson met the Azorian UFO investigator, Victor Suarez, and through him the Prof. Neythe Rodriguez de Abreu, all in Porto Alegre. Both investigators were taken by Jeferson's knowledge of the alien subjects and soon he was invited to attend conferences through Brazil with them. He met celebrated investigators like Prof. Irene Granchi and the General Alfredo Moacyr de Mendonça Uchoa. Later he worked for the (then) official Brazilian government news agency, EBN.

From his work in EBN in Brasilia, capital of Brasil, he went to Argentina to meet Fabio Zerpa, investigator and editor of the international magazine Cuarta Dimension (Fourth Dimension). For six months, he lived near Buenos Aires at the town of Lomas de Zamora. He learned quite a deal with Mr. Zerpa. Once back to Brazil in Porto Alegre, he was invited to attend conferences through Brazil and be the editor of Prof. Neythe's magazine named Disco Voador (Flying Saucer) for some time. Afterwards, he returned to Brasilia by invitation of General Uchoa for a more direct approach on research with his study group in an office in Brasilia and on his private ranch where strange events happened often.

After a time there, he was invited to attend a conference in Spain and to be a guest of the celebrated Catalonyan author and investigator, Antonio Ribera. Through General Alfredo Uchoa, Jeferson was invited to visit Colonel Wendelle C. Stevens in Tucson, Arizona, USA and became good friends with him.

Years later, Jeferson caught the attention of the U.S. Department of Defense and became part of three black budget projects: the Blue Planet Project, the Pulsar Project and Project Nemesis, all non-official projects ran by DARPA (Defense Advanced Research Projects Agency, at Arlington, Virginia and NSA (National Security Agency) at Ft. Mead, Maryland. He was chosen not so much for his knowledge but because such projects involved contactees and abductees. The things he saw and experienced there made quite an impression on him, forever.

To enter such projects is somehow simpler than leaving them. Without more details, he managed to exit them, or so he thought. He knew too much and killing him would only validate his claims or create a martyr. The devious things such agencies do are not to only contain the target but to destroy he or she's credibility, annihilating one's reputation. And thus, Jeferson went to a federal U.S. prison, not only losing his freedom but also his very identity as his name, country and history were changed by his foes. He became somebody else for decades, unjustly and illegally incarcerated under a name that wasn't his and a life that shouldn't have to be as it was. For decades, his Brazilian family looked for him but he was someone else. And he was told that his entire family died in a house fire. He was beaten and tortured to a point to be using a wheelchair for years.

He sent his private notes out to friends in Canada in ways similar to an action movie proving that reality can be even more amazing than fiction. It was how the books The Blue Planet Project and the Pulsar Project were written, and later on stolen from an insidious person that published the books without giving a penny to him. The person who stole Jeferson's manuscripts used lies and a silk tongue to take the book Jeferson wrote in prison, The Blue Planet Mandate. The unscrupulous man made several other books from Jeferson's notes and procured himself a great deal of money. Jeferson was left in prison with someone else claiming his work, but he never lost hope.

His alien allies guaranteed to him that it wasn't the end by far. And it was so. He found his family alive in Brazil and learned he had been lied to. In 2019, he unexpectedly was released from captivity and had his identity restored. As a miracle, he found his way home to Brazil after a long hiatus. And just several months later came the pandemic, COVID-19.

But when he was still in captivity in the U.S. another miracle happened to him. He found yet another family in the most unexpected ways and it was how A.B.O. COMIX became part of his life. If it wasn't for A.B.O. COMIX he wouldn't be able to survive the quarantine during COVID-19. And now marks his official return with this book.

www.ingramcontent.com/pod-product-compliance
Lightning Source LLC
Chambersburg PA
CBHW050730030426
42336CB00012B/1497